AMERICAN MATHEMATICAL SOCIETY
TRANSLATIONS
Series 2
Volume 129

Two Papers on Extremal Problems in Complex Analysis

by

S. Ya. Khavinson

American Mathematical Society
Providence, Rhode Island

Translated by D. KHAVINSON

Translation edited by LEV. J. LEIFMAN

1980 *Mathematics Subject Classification* (1985 *Revision*). Primary 30D55; Secondary 30E99.

Library of Congress Cataloging-in-Publication Data

Khavinson, S. IA.

Two papers on extremal problems in complex analysis.

(American Mathematical Society translations; ser. 2, v. 129)

Translation of: Osnovy teorii ėkstremal'nykh zadach dli͡a ogranichennykh analiticheskikh funkt͡siĭ i ikh razlichnykh obobshcheniĭ; Osnovy teorii ėkstremal'nykh zadach dli͡a ogranichennykh analiticheskikh funkt͡siĭ s dopolnitel'nymi uslovii͡ami.

Contents: Foundations of the theory of extremal problems for bounded analytic functions and various generalizations of them; Foundations of the theory of extremal problems for bounded analytic functions with additional conditions.

1. Functions of complex variables. 2. Extremal problems (Mathematics) I. Khavinson, S. I͡A. Osnovy teorii ėkstremal'nykh zadach dli͡a orgranichennykh analiticheskikh funkt͡siĭ s dopolnitel'nymi uslovii͡ami. English. 1986. II. Title. III. Series.

QA3.A572 ser. 2, vol. 129 510 s [515.9] 86-3545
[QA331]
ISBN 0-8218-3098-8

Contents

Russian Contents*

С. Я. Хавинсон, Основы теории экстремальных задач для ограниченных аналитических функций и их различных обобщений, Учебное пособие, Факульт. Повышения квалификации преподавателей, Инж.-строит. инст., Москва, 1981

С. Я. Хавинсон, Основы теории экстремальных задач для ограниченных аналитических функций с дополнительными условиями, Учебное пособие, Факульт. Повышения квалификации преподавателей, Инж.-строит. инст., Москва, 1981

*The American Mathematical Society scheme for transliteration of Cyrillic may be found at the end of index issues of *Mathematical Reviews*.

Amer. Math. Soc. Transl.
(2) Vol. **129**, 1986

Foundations of the Theory of Extremal Problems for Bounded Analytic Functions and Various Generalizations of Them*

S. YA. KHAVINSON

Introduction

These lecture notes are intended to provide a source for the students in the Department of Continuing Education for Instructors at the Moscow Institute of Civil Engineering who are studying additional chapters of the theory of analytic functions. The notes can be used for an advanced study of functional analysis and the theory of extremal problems. The main approach to extremal problems developed in these notes consists in the use of duality. The applications of duality methods in best approximation theory (where they originated in the important work of M. G. Kreĭn and S. M. Nikol'skiĭ) and in mathematical programming have been discussed in numerous monographs and textbooks. However, there has been no systematic account of the theory of analytic functions in this direction. Here, we present the duality method in application to the solution of a wide circle of extremal problems in various classes of analytic functions in multiply connected domains.

The author of these notes was the first who (since 1949) started applying the duality method to extremal problems in the theory of analytic functions. Now the number of research papers dealing with this question is large, and it seems useful to give a detailed survey of the main results. The section dedicated to this question in the well-known book by Duren [27] only touches on the very beginnings of this topic for functions in the unit disk. However, the most important results are obtained for multiply connected domains.

In conclusion we note that the bibliographical remarks and the list of references do not claim to be complete.

1980 *Mathematics Subject Classification* (1985 *Revision*). Primary 30D55; Secondary 30E99.

*Translation of a set of lecture notes published as a textbook (for teachers' continuing education), Moskov. Inzh.-Stroit Inst., Moscow, 1981.

§1. Duality of extremal problems in linear spaces

In this section, we give the simplest results concerning the duality relations between various extremal problems in normed spaces. The existence of such duality is helpful in more profound understanding of each one of the dual problems, and it is a starting point for establishing the characteristic properties of the extremal elements, for studying their uniqueness, and for establishing the various quantitative relations.

Usually, one of a pair of dual problems in the theory of analytic functions is a linear extremal problem, i.e., the problem of finding the extremum of a linear functional. The second problem is one of best approximation.

Let us agree on some notation. Let X be a normed space, real or complex, and let X^* be the dual space, i.e., the space of continuous linear functionals on X. Elements of X are denoted by f, ϕ, etc. (having in mind the following applications to analytic functions). Elements of X^* are denoted by l, λ, L, Λ, etc. If E is a subspace of X, then $E^\perp$ denotes the annihilator of E in X^*, i.e., the set of all linear functionals $l \in X^*$ such that $l(f) = 0$ for all $f \in E$. If $\mathscr{E}$ is a subspace of X^*, then $\mathscr{E}_\perp$ denotes the set of all elements $f \in X$ such that $l(f) = 0$ for each $l \in \mathscr{E}$. S stands for the unit sphere in the space X, and S^* is the unit sphere in X^*.

THEOREM 1.1. *Let E be a subspace of the normed space X and let $l_0 \in X^*$ be an arbitrary continuous linear functional on X. Then*

$$\sup_{\substack{f\in E \\ \|f\|\leqslant 1}} |l_0(f)| = \inf_{l\in E^\perp} \|l_0 - l\|. \tag{1.1}$$

Moreover, there exists a linear functional $l^* \in E^\perp$ for which the infimum in (1.1) *is attained.*

PROOF. The supremum on the left-hand side of (1.1) is simply the norm of the functional l_0 over the subspace E:

$$\sup_{\substack{f\in E \\ \|f\|\leqslant 1}} |l_0(f)| = \|l_0\|_E. \tag{1.2}$$

Since for any $l \in E^\perp$ the functional $l_0 - l$ coincides with l_0 on E, we have

$$\|l_0\|_E = \|l_0 - l\|_E \leqq \|l_0 - l\|. \tag{1.3}$$

(The last norm is the norm of the functional $l_0 - l$ over the whole space X.) Thus

$$\|l_0\|_E \leqslant \inf_{l\in E^\perp} \|l_0 - l\|. \tag{1.4}$$

On the other hand, by the Hahn-Banach theorem ([30], Chapter IV, §1.4° or [25], Chapter II, §3.10), there exists a linear functional L on X which coincides with l_0 on E:

$$l_0(f) = L(f), \quad f \in E, \tag{1.5}$$

and

$$\|L\| = \|l_0\|_E. \tag{1.6}$$

Denote by l^* the difference $l_0 - L$. Then for $f \in E$ we have $l^*(f) = l_0(f) - L(f) = 0$ and hence $l^* \in E^\perp$. Rewriting (1.6) in the form $\|l_0\|_E = \|l_0 - l^*\|$ and comparing with (1.4), we complete the proof of our theorem.

Theorem 1.1 is the simplest example of a duality theorem. The linear extremal problem of computing the norm of a linear functional on the subspace E is associated with a problem of the best approximation in the dual space: in X^* the best approximation of l_0 by functionals of the subspace $E^\perp$ is sought.

For brevity, we shall often write max (min) instead of sup (inf) when the supremum (infimum) is certainly attained. Interchanging the roles of the spaces X and X^* in Theorem 1.1, we obtain the following statement.

THEOREM 1.2. *Let E be a subspace of the normed space X and let ω be an arbitrary element of X. Then*

$$\max_{\substack{l \in E^\perp \\ \|l\| \leq 1}} |l(\omega)| = \inf_{\phi \in E} \|\omega - \phi\|. \tag{1.7}$$

PROOF. If $\omega \in \overline{E}$, the closure of E, then (1.7) reduces to the equality $0 = 0$. So we can assume that $\inf\|\omega - \phi\| = d > 0$. For any functional $l \in E^\perp$, $\|l\| \leqslant 1$, and any $\phi \in E$ we have

$$|l(\omega)| = |l(\omega - \phi)| \leqslant \|l\| \cdot \|\omega - \phi\| \leqslant \|\omega - \phi\|.$$

This implies the following inequality:

$$\sup_{\substack{l \in E^\perp \\ \|l\| < 1}} |l(\omega)| < \inf_{\phi \in E} \|\omega - \phi\|. \tag{1.8}$$

Now let us consider the linear span E_1 of the subspace E and the element ω. E_1 consists of elements of the form $\alpha\omega + \phi$, where α is a scalar and $\phi \in E$. Define a linear functional l_0 on E by

$$l_0(\omega) - d = \inf_{\phi \in E} \|\omega - \phi\|, \qquad l(\phi) = 0, \forall \phi \in E \tag{1.9}$$

Let us find the norm of this functional on E_1. On one hand,

$$d = l_0(\omega) = l_0(\omega - \phi) \leqslant \|l_0\|_{E_1} \cdot \|\omega - \phi\|,$$

and since $\phi \in E$ is arbitrary, it follows that

$$d \in \|l_0\|_{E_1} \cdot \inf_{\phi \in E} \|\omega - \phi\| \leqslant \|l_0\|_{E_1} \cdot d.$$

Therefore, $\|l_0\|_{E_1} \geqslant 1$. On the other hand, if an element $f \in E_1$ has $\|f\| = 1$, then representing f in the form $f = \alpha\omega + \phi$, $\phi \in E$, we find that

$$1 = \|f\| = \|\alpha\omega + \phi\| + |\alpha| \cdot \|\omega + \phi/\alpha\| \geqslant |\alpha| \cdot d. \tag{1.10}$$

Hence

$$\|l_0\|_{E_1} = \sup_{\|f\| \leqslant 1} |l_0(f)| = \sup_{\|f\| \leqslant 1} \frac{|\alpha| \cdot d}{\|\alpha\omega + \phi\|} \leqslant \frac{|\alpha| \cdot d}{|\alpha| \cdot d} = 1.$$

Thus,

$$\|l_0\|_{E_1} = 1. \tag{1.11}$$

Extending l_0 without increasing its norm to the whole space X by the Hahn-Banach theorem, we can assume that

$$\|l_0\| = 1. \tag{1.12}$$

Since by (1.9)

$$l_0(\omega) = \inf_{\phi \in E} \|\omega - \phi\|, \tag{1.13}$$

combining this equality with (1.8) and (1.12), we complete the proof of the theorem.

One can restate Theorem 1.2 in a slightly different form. Let $E^*(\omega)$ be the set of functionals $l \in X^*$ such that

$$\operatorname{Re} l(\omega) = 1, \qquad l(\phi) = 0,\ \phi \in E \tag{1.14}$$

(we assume that $\omega \notin \overline{E}$). Then, under the hypothesis of Theorem 1.2,

$$\max_{l \in E^*(\omega)} \frac{1}{\|l\|} = \inf_{\phi \in E} \|\omega - \phi\|. \tag{1.15}$$

This follows immediately from (1.7) if the following facts are taken into account. For $l \in E^*(\omega)$ the functional $\lambda = l/\|l\| \in E^{\perp}$ and $\|\lambda\| = 1$; conversely, if $\lambda \in E^{\perp}$ and $\|\lambda\| = 1$, then $l = \lambda/\operatorname{Re}\lambda(\omega) \in E^*(\omega)$.

The equality (1.15) would not change if we slightly varied the definition of $E^*(\omega)$, including in it the functionals l such that

$$l(\omega) = 1, \qquad l(\phi) = 0, \quad \phi \in E. \tag{1.16}$$

Similar remarks can be made concerning Theorem 1.1. We leave it to the reader to work out the corresponding relations.

§2. Duality of extremal problems in the class of bounded analytic functions and in the classes E_p, $p \geqslant 1$

Let G be an n-connected domain with the boundary Γ consisting of the rectifiable contours $\gamma_1, \dots, \gamma_n$. To avoid diffuseness, for now we assume that G is finite. Let $\rho(\zeta)$ be a continuous function on Γ, with

$$\rho(\zeta) > 0. \tag{2.1}$$

Consider the class $C^1_{\rho(\zeta)}$ of functions analytic in G, continuous in the closure $\overline{G}$ of G and satisfying the followng inequality on the boundary:

$$|f(\zeta)| \leqslant \rho(\zeta), \qquad \zeta \in \Gamma. \tag{2.2}$$

Let $C(\Gamma)$ denote, as usual, the space of continuous complex-valued functions on Γ. Define the subspace in $C(\Gamma)$ consisting of functions $\mathscr{F}(\zeta)$ of the form

$$\mathscr{F}(\zeta) = f(\zeta)/\rho(\zeta), \tag{2.3}$$

where $f(\zeta)$ denotes boundary values on Γ of a function analytic in G and continuous in $\overline{G}$. We denote this subspace by $C_A(\Gamma)$. When the function $f(\zeta)$ runs over the class $C^1_{\rho(\zeta)}$, the function $\mathscr{F}(\zeta)$ defined by (2.3) runs over the unit sphere $C^1_A(\Gamma)$ in the subspace $C_A(\Gamma)$.

To make use of Theorem 1.1, we shall determine the form of linear functionals orthogonal to the subspace $C_A(\Gamma)$. Let a functional $l(\mathscr{F})$ on $C(\Gamma)$ be orthogonal to $C_A(\Gamma)$. If μ is a (complex) Baire measure representing the functional l, then we have

$$l(\mathscr{F}) = \int_\Gamma \mathscr{F}(\zeta)\, d\mu = \int_\Gamma f(\zeta)\, d\mu_1,$$

where the measure $d\mu_1 = d\mu/\rho(\zeta)$, and $f = \mathscr{F}\rho$. Taking for $f(\zeta)$ the function $(\zeta - z)^{-1}$ as z varies in the whole complement of $\overline{G}$, we obtain the identity

$$\int_\Gamma \frac{d\mu_1}{\zeta - z} \equiv 0, \qquad z \notin \overline{G}. \tag{2.4}$$

According to the F. and M. Riesz theorem (see the Appendix, Theorem 3), we may conclude that the measure μ_1 is absolutely continuous and $d\mu_1 = \phi(\zeta)\, d\zeta$, where $\phi(\zeta)$ are boundary values on Γ of a certain analytic function of the class $E_1(G)$ (see the Appendix). Hence, the measure $d\mu$ orthogonal to $C_A(\Gamma)$ has the following form:

$$d\mu = \rho(\zeta)\phi(\zeta)\, d\zeta, \qquad \phi(z) \in E_1(G). \tag{2.5}$$

Conversely, any measure μ which has the form (2.5) is orthogonal to $C_A(\Gamma)$. If $\mathscr{F}(\zeta) \in C_A(\Gamma)$, then

$$\int_\Gamma \mathscr{F}(\zeta)\rho(\zeta)\phi(\zeta)\, d\zeta = \int_\Gamma f(\zeta)\phi(\zeta)\, d\zeta.$$

Indeed $f(z)\phi(z) \in E_1(G)$, so the Cauchy theorem holds for its boundary values.

Now we are able to state the following duality theorem.

THEOREM 2.1. *Let g_0 be an arbitrary complex Baire measure on Γ and let $\rho(\zeta) > 0$ be continuous on Γ. Then*

$$\sup_{f \in C^1_{\rho(\zeta)}} \left| \int_\Gamma f(\zeta)\, dg_0 \right| = \min_{\phi \in E_1(G)} \int_\Gamma \rho(\zeta) |dg_0 - d\nu|, \tag{2.6}$$

where $d\nu = \phi(\zeta)\, d\zeta$.

PROOF. We will use Theorem 1.1. In our case $E = C_A(\Gamma)$, and $E^{\perp}$ is described by the measures (2.5). We obtain

$$\sup_{f\in C^1_{\rho(\zeta)}}\left|\int_\Gamma f(\zeta)\,dg_0\right| = \sup_{\mathscr{F}\in C^1_A(\Gamma)}\left|\int_\Gamma \mathscr{F}(\zeta)\rho(\zeta)\,dg_0\right|$$

$$= \inf_{\mu\in C_A(\Gamma)^{\perp}}\int_\Gamma |\rho(\zeta)\,dg_0 - d\mu|$$

$$= \inf_{\phi\in E_1(G)}\int_\Gamma |\rho(\zeta)\,dg_0 - \rho(\zeta)\phi(\zeta)\,d\zeta|$$

$$= \inf_{\phi\in E_1(G)}\int_\Gamma \rho(\zeta)|dg_0 - d\nu|,$$

where $d\nu = \phi(\zeta)\,d\zeta$. Moreover, from Theorem 1.1 it follows that the infimum in (2.6) is attained for a certain $\phi^*(z)\in E_1(G)$. The theorem is proved.

Let us now denote by $B^1_{\rho(\zeta)}$ the class of bounded analytic functions in G satisfying (2.2) almost everywhere on Γ ($\rho(\zeta)$ as above), i.e.

$$B^1_{\rho(\zeta)} = \{f(z);\, f(z)\in B(G),\, |f(\zeta)|/\rho(\zeta)\leqslant 1 \text{ a.e. on } \Gamma\}, \tag{2.7}$$

where $B(G) = H_\infty(G)$ is the space of bounded analytic functions in G. If $\rho(\zeta)\equiv 1$, we denote the corresponding class $B^1_{\rho(\zeta)}$ by $B^1(G)$. $B^1(G)$ is the unit sphere in $B(G)$.

THEOREM 2.2. *For any summable function* $\omega(\zeta)$ *on* Γ,

$$\sup_{f\in C^1_{\rho(\zeta)}}\left|\int_\Gamma f(\zeta)\omega(\zeta)\,d\zeta\right| = \sup_{f\in B^1_{\rho(\zeta)}}\left|\int_\Gamma f(\zeta)\omega(\zeta)\,d\zeta\right|$$

$$= \min_{\phi\in E_1(G)}\int_\Gamma \rho(\zeta)|\omega(\zeta)-\phi(\zeta)|\cdot|d\zeta|. \tag{2.8}$$

PROOF. The equality between the last two terms in (2.8) follows from the previous theorem. It is also clear that the intermediate term in (2.8) is not less than the first one, since $B^1_{\rho(\zeta)}\supset C^1_{\rho(\zeta)}$. On the other hand, for any $f(z)\in B^1_{\rho(\zeta)}$ and any $\phi(z)\in E_1(G)$ the product $f(z)\phi(z)\in E_1(G)$ and, therefore, $\int_\Gamma f(\zeta)\phi(\zeta)\,d\zeta = 0$. Hence,

$$\left|\int_\Gamma f(\zeta)\omega(\zeta)\,d\zeta\right| = \left|\int_\Gamma f(\zeta)[\omega(\zeta)-\phi(\zeta)]\,d\zeta\right|$$

$$\leqslant \int_\Gamma |f(\zeta)||\omega(\zeta)-\phi(\zeta)||d\zeta|$$

$$\leqslant \int_\Gamma \rho(\zeta)|\omega(\zeta)-\phi(\zeta)||d\zeta|. \tag{2.9}$$

Thus,

$$\sup_{f\in B^1_{\rho(\zeta)}}\left|\int_\Gamma f(\zeta)\omega(\zeta)\,d\zeta\right| \leqslant \inf_{\phi\in E_1(G)}\int_\Gamma \rho(\zeta)|\omega(\zeta)-\phi(\zeta)||d\zeta|. \quad (2.10)$$

The inequality (2.10) together with the equality of the last two terms in (2.8) mentioned above completes the proof of (2.8) in full.

Now let $\rho(\zeta)$ be a measurable function on Γ satisfying the conditions

$$0 < m \leqslant \rho(\zeta) \leqslant M < +\infty. \quad (2.11)$$

Denote by $E^1_{\rho(\zeta)}$ a subclass of $E_1(G)$ consisting of functions whose boundary values satisfy the condition

$$\int_\Gamma \frac{|f(\zeta)|}{\rho(\zeta)}\,ds \leqslant 1. \quad (2.12)$$

THEOREM 2.3. *For any bounded function* $\omega(\zeta)$ *on* Γ,

$$\sup_{f\in E^1_{\rho(\zeta)}}\left|\int_\Gamma f(\zeta)\omega(\zeta)\,d\zeta\right| = \min_{\phi\in B(G)} \operatorname{ess\,sup}\left[\rho(\zeta)|\omega(\zeta)-\phi(\zeta)|\right]. \quad (2.13)$$

PROOF. In the space $L_1(\Gamma)$ of all summable functions on Γ, consider the subspace E consisting of functions $\mathscr{F}(\zeta)$ of the form

$$\mathscr{F}(\zeta) = f(\zeta)/\rho(\zeta), \qquad f(z)\in E_1(G). \quad (2.14)$$

When $\mathscr{F}(\zeta)$ runs over the unit sphere E^1 of E, the function $f(z)$ in (2.14) runs over the class $E^1_{\rho(\zeta)}$.

Let us find the annihilator of the subspace E in the space $L_1(\Gamma)^* = L_\infty(\Gamma)$. Let the functional

$$l(\mathscr{F}) = \int_\Gamma \mathscr{F}(\zeta)\psi(\zeta)\,d\zeta, \quad (2.15)$$

$\psi(\zeta)\in L_\infty(\Gamma)$, be orthogonal to E. This means that

$$\int_\Gamma \mathscr{F}(\zeta)\psi(\zeta)\,d\zeta = \int_\Gamma f(\zeta)\frac{\psi(\zeta)}{\rho(\zeta)}\,d\zeta = 0, \qquad f\in E_1(G).$$

Taking as $f(\zeta)$ the function $(\zeta - z)^{-1}$, $z\notin \overline{G}$, we obtain

$$\int_\Gamma \frac{\psi(\zeta)\,d\zeta}{\rho(\zeta)(\zeta - z)} \equiv 0, \qquad z\notin\overline{G}. \quad (2.16)$$

From this it follows immediately that $\psi(\zeta)/\rho(\zeta)$ is the boundary value of an analytic function in G of the class $E_1(G)$. Now, if G is a Smirnov domain ($G\in S$) (see the Appendix), then according to the Smirnov theorem (Theorem 2 in the Appendix) the function $\psi(\zeta)/\rho(\zeta)$, which belongs to the class $E_1(G)$ and is bounded on Γ, is the boundary value of the function $\phi(z)\in B(G)$:

$$\psi(\zeta) = \rho(\zeta)\phi(\zeta), \qquad \phi(z)\in B(G). \quad (2.17)$$

It is more difficult to justify (2.17) without the assumption that $G \in S$. Let $w = \alpha(z)$ be a conformal mapping of G onto the canonical circular domain. Then $\alpha'(z) \in E_1(G)$ (see the Appendix, Theorem 1), and the functions $\alpha'(\zeta)/(\zeta - z)$, $z \notin \overline{G}$, also belong to $E_1(G)$. Therefore, together with (2.16), we can write down the following orthogonality conditions:

$$\int_\Gamma \frac{\alpha'(\zeta)\psi(\zeta)\,d\zeta}{(\zeta - z)\rho(\zeta)} \equiv 0, \qquad z \notin \overline{G}. \tag{2.18}$$

From this it follows that

$$Q(\zeta) = \frac{\alpha'(\zeta)\psi(\zeta)}{\rho(\zeta)} \in E_1(G).$$

But then

$$\phi(\zeta) = Q(\zeta)/\alpha'(\zeta) = \psi(\zeta)/\rho(\zeta) \in H_1(G).$$

The function $\phi(z) \in H_1(G)$ is representable by the Green integral of its boundary values (see the Appendix). Since its boundary values are bounded, we conclude that $\phi(z) \in B(G)$. We have justified (2.17) again, this time without the assumption that $G \in S$. Now, from (1.1), we find

$$\begin{aligned}\sup_{f \in E^1_{\rho(\zeta)}} \left| \int_\Gamma f(\zeta)\omega(\zeta)\,d\zeta \right| &= \sup_{\mathscr{F} \in E^1} \left| \int_\Gamma \mathscr{F}(\zeta)\rho(\zeta)\omega(\zeta)\,d\zeta \right| \\ &= \min_{\psi \in E^\perp} \operatorname*{ess\,sup}_{\zeta \in \Gamma} |\rho(\zeta)\omega(\zeta) - \psi(\zeta)| \\ &= \min_{\phi \in B(G)} \operatorname*{ess\,sup}_{\zeta \in \Gamma} \left[\rho(\zeta)|\omega(\zeta) - \phi(\zeta)|\right],\end{aligned}$$

Q.E.D.

Let $\rho(\zeta)$ satisfy (2.11). Introduce the classes $E^1_{\rho(\zeta),q}$, $q \geqslant 1$. The class $E^1_{\rho(\zeta),q}$ consists of boundary values of the functions $f(z) \in E_q(G)$ (see the Appendix) for which

$$\int_\Gamma \left| \frac{f(\zeta)}{\rho(\zeta)} \right|^q ds \leqslant 1. \tag{2.19}$$

In the case $q = 1$, we get the class $E^1_{\rho(\zeta)}$ considered above. For $\rho(\zeta) \equiv 1$, the corresponding classes are denoted by $E^1_{\rho(\zeta),q} = E^1_q$, $q \geqslant 1$. Those are simply the unit spheres in $E_q(G)$.

THEOREM 2.4. *Let $q > 1$ and $w(\zeta) \in L_p(\Gamma)$, $1/p + 1/q = 1$. Then the following duality relation holds*:

$$\sup_{f \in E^1_{\rho(\zeta),q}} \left| \int_\Gamma f(\zeta)\omega(\zeta)\,d\zeta \right| = \min_{\phi \in E_p(G)} \left\{ \int_\Gamma \left[\rho(\zeta)|\omega(\zeta) - \phi(\zeta)|\right]^p ds \right\}^{1/p}. \tag{2.20}$$

The proof differs very little from the proof of Theorem 2.3, and we omit it.

It is possible to strengthen considerably the above duality relations by getting rid of the requirement of continuity of $\rho(\zeta)$ in (2.8) and weakening the constraints (2.11) on $\rho(\zeta)$ in (2.13) and (2.20).

§3. Bounded analytic functions. Existence and uniqueness of extremal functions

LEMMA 3.1. *The class $B^1_{\rho(\zeta)}$ is compact with respect to convergence inside (i.e., on compact subsets of) the domain G.*

PROOF. If $f(z) \in B^1_{\rho(\zeta)}$, then

$$\sup_{z\in G} |f(z)| = \operatorname*{ess\,sup}_{\zeta\in\Gamma} |f(\zeta)| \leqslant \rho(\zeta).$$

Therefore, all functions of the class $B^1_{\rho(\zeta)}$ are uniformly bounded.

We can choose a subsequence from the sequence $\{f_n(z)\} \subset B^1_{\rho(\zeta)}$ which converges uniformly inside G to a bounded function $f(z)$ (to preserve the notation we assume that $\{f_n(z)\}$ itself converges to $f(z)$). We must check that $f(z) \in B^1_{\rho(\zeta)}$. In view of weak convergence of the boundary values $\{f_n(\zeta)\}$ to $f(\zeta)$ (see the Appendix, Theorem 4) we obtain

$$\lim_{n\to\infty} \int_\Gamma f_n(\zeta)\omega(\zeta)\,d\zeta = \int_\Gamma f(\zeta)\omega(\zeta)\,d\zeta$$

for any summable function $\omega(\zeta)$ on Γ. Setting $\mathscr{F}_n(\zeta) = f_n(\zeta)/\rho(\zeta)$ and $\mathscr{F}(\zeta) = f(\zeta)/\rho(\zeta)$, we see that the sequence $\mathscr{F}_n(\zeta)$ converges weakly in the space $L_1(\Gamma)^* = L_\infty(\Gamma)$ to $\mathscr{F}(\zeta)$. Therefore, $\|\mathscr{F}\|_\infty \leqslant \underline{\lim}_{n\to\infty}\|\mathscr{F}_n\| \leqslant 1$, and so $|f(\zeta)|/\rho(\zeta) \leqslant 1$, Q.E.D.

THEOREM 3.2. *Let $\omega(\zeta)$ be an arbitrary summable function on Γ (we exclude the trivial case $\omega(\zeta) \in E_1$). Let $\rho(\zeta) > 0$ be continuous on Γ. Then the following assertions are true:*

1. *There exist extremal functions $f^*(z) \in B^1_{\rho(\zeta)}$ and $\phi^*(z) \in E_1(G)$ in (2.8).*

2. *The functions $f^*(z) \in B^1_{\rho(\zeta)}$ and $\phi^*(z) \in E_1(G)$ are extremal in (2.8) if and only if*

$$f^*(\zeta)[\omega(\zeta) - \phi^*(\zeta)]\,d\zeta = e^{i\delta}\cdot \rho(\zeta)|\omega(\zeta) - \phi^*(\zeta)|\,ds \tag{3.1}$$

almost everywhere on Γ, where δ is a real constant.

3. *An extremal function $f^*(z)$ is unique up to a factor $e^{i\alpha}$. If the domain G is simply connected, then an extremal function $\phi^*(z)$ is unique. If G is n-connected $(n > 1)$, then an extremal function $\phi^*(z)$ is unique provided that $f^*(z)$ has more than $n-2$ zeros in G or that on a certain set $T \subset \Gamma$, $\operatorname{meas} T > 0$, $|f^*(\zeta)| < \rho(\zeta)$. Otherwise, an extremal function $\phi^*(z)$ may fail to be unique.*

PROOF. Existence of an extremal function $\phi^*(z)$ has already been proved. Let us prove the existence of an extremal function $f^*(z) \in B^1_{\rho(\zeta)}$. Let

$$\lambda = \sup_{f\in B^1_{\rho(\zeta)}} \left|\int_\Gamma f(\zeta)\omega(\zeta)\,d\zeta\right|.$$

Take a sequence $\{f_n(z)\} \subset B^1_{\rho(\zeta)}$ such that

$$\lim_{n\to\infty}\left|\int_\Gamma f_n(\zeta)\omega(\zeta)\,d\zeta\right| = \lambda.$$

In view of the compactness of $B^1_{\rho(\zeta)}$ we can choose a subsequence of $\{f_n(z)\}$ (we will not change the notation) converging uniformly on compact subsets of G to the function $f^*(z) \in B^1_{\rho(\zeta)}$. Then, by the theorem on weak convergence of boundary values (Appendix, Theorem 4),

$$\int_\Gamma f^*(\zeta)\omega(\zeta)\,d\zeta = \lim_{n\to\infty}\int_\Gamma f_n(\zeta)\omega(\zeta)\,d\zeta.$$

Thus, we have constructed an extremal function $f^*(z)$. Now, we have

$$\begin{aligned}\left|\int_\Gamma f^*(\zeta)\omega(\zeta)\,d\zeta\right| &= \left|\int_\Gamma f^*(\zeta)[\omega(\zeta)-\phi^*(\zeta)]\,d\zeta\right| \\ &\leqslant \int_\Gamma \rho(\zeta)|\omega(\zeta)-\phi^*(\zeta)|\,ds.\end{aligned}$$

Since $f^*(z)$ and $\phi^*(z)$ are extremal functions, in our chain of relations we must have equalities everywhere, according to (2.8). This implies that almost everywhere on a set $A = \{\zeta \in \Gamma\colon \omega(\zeta) \neq \phi^*(\zeta)\}$ we have

$$|f^*(\zeta)| = \rho(\zeta),$$
$$\arg\{f^*(\zeta)[\omega(\zeta)-\phi^*(\zeta)]\,d\zeta\} = \text{const} = \delta.$$

Now it is clear that

$$f^*(\zeta) = [\omega(\zeta)-\phi^*(\zeta)]\,d\zeta = e^{i\delta}\cdot\rho(\zeta)|\omega(\zeta)-\phi^*(\zeta)|\,ds$$

on A. But the last relation is trivial on the set where $\omega(\zeta) = \phi^*(\zeta)$. (3.1) is proved.

Conversely, let (3.1) hold almost everywhere on Γ for two functions $f^*(z) \in B^1_{\rho(\zeta)}$ and $\phi^*(z) \in E_1(G)$. Then

$$\begin{aligned}\left|\int_\Gamma f^*(\zeta)\omega(\zeta)\,d\zeta\right| &= \left|\int_\Gamma f^*(\zeta)[\omega(\zeta)-\phi^*(\zeta)]\,d\zeta\right| \\ &= \int_\Gamma \rho(\zeta)|\omega(\zeta)-\phi^*(\zeta)|\,ds,\end{aligned}$$

and in view of (2.8) it is clear that $f^*(z)$ realizes

$$\sup_{f\in B^1_{\rho(\zeta)}}\left|\int_\Gamma f(\zeta)\omega(\zeta)\,d\zeta\right|$$

and $\phi^*(z)$ realizes

$$\inf_{\phi\in E_1(G)}\int_\Gamma \rho(\zeta)|\omega(\zeta)-\phi(\zeta)|\,ds.$$

The proof of assertion 2 is completed.

Let us investigate the uniqueness question for extremal functions. Let $f_1^*(z)$ and $f_2^*(z)$ be extremal functions for which the supremum in (2.8) is attained. According to (3.1) we have

$$f_1^*(\zeta)[\omega(\zeta)-\phi^*(\zeta)]\,d\zeta = e^{i\delta_1}\rho(\zeta)|\omega(\zeta)-\phi^*(\zeta)|\,ds;$$
$$f_2^*(\zeta)[\omega(\zeta)-\phi^*(\zeta)]\,d\zeta = e^{i\delta_2}\rho(\zeta)|\omega(\zeta)-\phi^*(\zeta)|\,ds.$$

(We take the same function $\phi^*(z)$ in both equalities.) But on the set A, where $\omega(\zeta)\neq\phi^*(\zeta)$, we have

$$f_1^*(\zeta) = e^{i(\delta_1-\delta_2)}\cdot f_2^*(\zeta).$$

Since meas $A>0$, according to the Luzin-Privalov uniqueness theorem, we obtain

$$f_1^*(z) \equiv e^{i(\delta_1-\delta_2)}\cdot f_2^*(z).$$

On the other hand, if $f^*(z)$ is a certain extremal function for the problem of

$$\sup_{f\in B^1_{\rho(\zeta)}}\left|\int_\Gamma f(\zeta)\omega(\zeta)\,d\zeta\right|,$$

then each function $e^{i\delta}f^*(z)$ is also an extremal function. This proves that the extremal function $f^*(z)$ is defined up to a factor $e^{i\delta}$.

Now we proceed to the uniqueness problem for $\phi^*(z)$. Assume that there exist two extremal functions, $\phi_1^*(z)$ and $\phi_2^*(z)$. Put $Q(z)=\phi_1^*(z)-\phi_2^*(z)$. Take the extremal function $f^*(z)\in B^1_{\rho(\zeta)}$ for which

$$\int_\Gamma f^*(\zeta)\omega(\zeta)\,d\zeta>0.$$

Then (again in view of (3.1)), almost everywhere on Γ we have

$$f^*(\zeta)[\omega(\zeta)-\phi_1^*(\zeta)]\,d\zeta = \rho(\zeta)|\omega(\zeta)-\phi_1^*(\zeta)|\,ds;$$
$$f^*(\zeta)[\omega(\zeta)-\phi_2^*(\zeta)]\,d\zeta = \rho(\zeta)|\omega(\zeta)-\phi_2^*(\zeta)|\,ds.$$

Subtracting the first equality from the second, we find that $f^*(\zeta)Q(\zeta)\,d\zeta$ is real-valued almost everywhere on Γ.

First, let G be simply connected and let $z=\alpha(t)$ be a conformal mapping of the unit disk $|t|<1$ onto G. As $f^*(z)\in B(G)$ and $\phi^*(z)\in E_1(G)$, we see that $f^*(z)Q(z)\in E_1(G)$. Also, almost everywhere on the circle $|t|=1$ we obtain

$$\Phi[\alpha(t)]\alpha'(t)\,dt - \Phi(\zeta)\,d\zeta \qquad (\Phi(z)=f^*(z)(Q(z))).$$

Therefore, $\Phi[\alpha(t)]\alpha'(t)\,dt$ is real-valued almost everywhere on $|t|=1$. Hence, the function $i\Phi[\alpha(t)]\alpha'(t)t$, which belongs to the class H_1, has real-valued boundary values almost everywhere on $|t|=1$. Representing this function by the Poisson integral, we observe that

$$\operatorname{Im}[i\Phi[\alpha(t)]\alpha'(t)t]\equiv 0.$$

Therefore $i\Phi[\alpha(t)]\alpha'(t)t\equiv$ const. But the last expression is equal to zero at $t=0$. Thus,

$$\Phi[\alpha(t)]\alpha'(t)t\equiv 0.$$

Hence $\Phi[\alpha(t)]\equiv 0$ and $\Phi(z)=f^*(z)Q(z)\equiv 0$.

Since $f^*(z) \not\equiv 0$, we obtain that $Q(z) \equiv 0$ and $\phi_1^*(z) \equiv \phi_2^*(z)$. Consider the case of a multiply connected domain G. The above argument for a simply connected domain reduces to proving that if for a certain function $\Phi(z) \in E_1$ the equality $\operatorname{Im}(\Phi(\zeta)\,d\zeta) = 0$ holds almost everywhere on the boundary, then $\Phi(z) \equiv 0$. This is not true for multiply connected domains, as follows from the example given below.

Let $f^*(z)$ have more than $n-2$ zeros. Again writing $\Phi(z) = f^*(z)Q(z)$, we have

$$\operatorname{Im}(\Phi(\zeta)\,d\zeta) = 0$$

almost everywhere on Γ. Our domain G can always be mapped conformally onto a domain G_1 of the t-plane bounded by regular analytic contours. Denote by Γ_1 the boundary of G_1. Let the function $z = \alpha(t)$ give the above conformal mapping of G_1 onto G. Since $\Phi(z) \in E_1$ in G, it easily follows that $\Phi[\alpha(t)]\alpha'(t) \in E_1(G_1)$ and

$$\operatorname{Im}[\Phi(\alpha(t))\alpha'(t)\,dt] = 0$$

almost everywhere on Γ_1. From the last equality and the symmetry principle generalized to the class E_1, it follows that the function $R(t) = \Phi(\alpha(t))\alpha'(t)$ is analytic in $\overline{G}_1$. (The properties of functions of the class E_1 allow us to extend the classical proof of the symmetry principle.) Let $R(t)$ have k zeros in G_1 and ν zeros on Γ_1 (counting multiplicities). By the argument principle

$$k + \frac{\nu}{2} = \frac{1}{2\pi}\sum_{S_i} \Delta_{S_i} \arg R(t). \tag{3.2}$$

Here, S_i is an open arc between two neighboring zeros on Γ_i; $\Delta_{S_i} \arg R(t)$ is an increment of the argument of $R(t)$ along this arc. The sum is taken over all arcs S_i. For our function $R(t)$ we obtain

$$\Delta_{S_i} \arg R(t) = \Delta_{S_i} \arg \frac{R(t)\,dt}{dt} = -\Delta_{S_i} \arg dt.$$

Thus,

$$k + \frac{\nu}{2} = -\frac{1}{2\pi}\Delta_{\Gamma_1} \arg dt = n - 2$$

($\Delta_{\Gamma_1} \arg dt$ is the total increment of $\arg dt$ along the boundary). Therefore, the number of interior zeros of $R(t)$ is at most $n-2$. Recalling that

$$R(t) = f^*[\alpha(t)]Q[\alpha(t)]\alpha'(t)$$

and that $f^*(z)$ has more than $n-2$ zeros by our assumption, we arrive at the contradiction. This proves uniqueness of the extremal function $\phi^*(z)$.

Let us now assume that $|f^*(\zeta)| < \rho(\zeta)$ on the set $T \subset \Gamma$, $\operatorname{meas} T > 0$. From (3.1) (since it holds for all extremal functions $\phi^*(\zeta)$) it follows that $\omega(\zeta) = \phi^*(\zeta)$ on T. If this were not true, then the modulus of $f^*(\zeta)$ would have been equal to $\rho(\zeta)$ almost everywhere on T. So, all extremal functions coincide with $\omega(\zeta)$ on T, and so they are all equal on T. Being equal on a set of positive measure, they must be identically equal.

Now, let us construct an example where an extremal function $f^*(z)$ has less than $n-2$ zeros and an extremal function $\phi^*(z)$ is not unique. This will complete the proof of Theorem 3.2.

Let the boundary Γ of the domain G consist of analytic contours and let $\rho \equiv 1$. Let $\omega(\zeta) = ds/d\zeta$. We have

$$1 \cdot \omega(\zeta)\, d\zeta = |\omega(\zeta)|\, ds$$

almost everywhere. This implies that $f^*(z) \equiv 1$ is an extremal function for

$$\sup_{f \in B^1} \left| \int_\Gamma f(\zeta)\omega(\zeta)\, d\zeta \right|$$

and $\phi^*(z) \equiv 0$ is an extremal function for

$$\inf_{\phi \in E_1} \int_\Gamma |\omega(\zeta) - \phi(\zeta)|\, ds.$$

Let $\omega_k(z)$ be a harmonic function in G equal to one on γ_k and zero on γ_i, $i \neq k$ ($\omega_k(z)$ is the harmonic measure of the contour γ_k). Denote by $\omega_k^*(z)$ the harmonic conjugate of $\omega_k(z)$ and set

$$w_k(z) = \omega_k(z) + i\omega_k^*(z). \tag{3.3}$$

The function $w_k(z)$ is analytic in G and, as easily follows from the Cauchy-Riemann equations, $\operatorname{Im}[iw_k'(\zeta)\, d\zeta] = 0$ on Γ. Take an arbitrary linear combination

$$w(z) i \sum_1^n \lambda_k w_k'(z)$$

with real λ_k. Choose λ_k such that $|w(\zeta)| \leqslant q < 1 = |\omega(\zeta)|$ on Γ. Then, since $|w(\zeta)| < |\omega(\zeta)|$,

$$\arg\{[\omega(\zeta) - w(\zeta)]\, d\zeta\} = \arg\{\omega(\zeta)\, d\zeta\}.$$

Hence,

$$i[\omega(\zeta) - w(\zeta)]\, d\zeta = |\omega(\zeta) - w(\zeta)|\, ds.$$

Therefore, each such linear combination $w(z)$ is an extremal function for the problem of

$$\inf_{\phi \in E_1} \int_\Gamma |\omega(\zeta) - \phi(\zeta)|\, ds$$

as soon as $|w(\zeta)| < 1$.

§4. Classes E_p, $p \geqslant 1$. Existence and uniqueness of extremal functions

LEMMA 4.1. *If $\rho(\zeta) > 0$ is continuous, then the class $E^1_{\rho(\zeta)}$ is compact with respect to convergence inside (i.e., on compact subsets of) G.*

PROOF. From the inequality

$$\int_\Gamma \frac{|f(\zeta)|}{\rho(\zeta)}\, ds \leqslant 1,$$

defining the class $E^1_{\rho(\zeta)}$, it follows immediately that the functions of $E^1_{\rho(\zeta)}$ are uniformly bounded on the compact subsets of G. Therefore, we can choose a subsequence of the sequence $\{f_n(z)\} \subset E^1_{\rho(\zeta)}$ which converges uniformly on compact subsets of G. We assume that the sequence $\{f_n(z)\}$ itself already converges uniformly to $f(z)$ inside G. We shall show that $f(z) \in E^1_{\rho(\zeta)}$. According to the theorem on weak convergence of the boundary values of functions in E_1 (Theorem 4 in the Appendix), from the uniform convergence of $\{f_n(z)\}$ on compact subsets of G it follows that

$$\int_\Gamma f(\zeta)\omega(\zeta)\,d\zeta = \lim_{n\to\infty}\int_\Gamma f_n(\zeta)\omega(\zeta)\,d\zeta$$

for any continuous function $\omega(\zeta)$.

Replacing $f(\zeta)$ by $\mathscr{F}(\zeta) = f(\zeta)/\rho(\zeta)$ and $f_n(\zeta)$ by $\mathscr{F}_n(\zeta) = f_n(\zeta)/\rho(\zeta)$, we obtain

$$\int_\Gamma \mathscr{F}(\zeta)\omega(\zeta)\,d\zeta = \lim_{n\to\infty}\int_\Gamma \mathscr{F}_n(\zeta)\omega(\zeta)\,d\zeta$$

for any continuous $\omega(\zeta)$. Therefore, by the general theorem on norms of a weakly converging sequence of functionals (in our case over the space $C(\Gamma)$), we have

$$\int_\Gamma |\mathscr{F}(\zeta)|\,ds \leqslant \varliminf_{n\to\infty}\int_\Gamma |\mathscr{F}_n(\zeta)|\,ds \leqslant 1,$$

Q.E.D.

THEOREM 4.2. *Let the hypothesis of Theorem* 3.2 *be satisfied* (*we exclude the trivial case* $\omega(\zeta) \in B(G)$). *Then the following assertions are true*:

1. *There exists an extremal function* $\phi^*(z)$ *for which the infimum on the right-hand side of* (2.13) *is attained. If* $\omega(\zeta)$ *and* $\rho(\zeta)$ *are continuous on* Γ, *then there exists an extremal function* $f^*(z)$ *for which the supremum on the left-hand side of* (2.13) *is attained. If* $\omega(\zeta)$ *is discontinuous, then an extremal function* $f^*(z)$ *need not exist.*

2. *The functions* $f^*(z) \in E^1_{\rho(\zeta)}$ *and* $\phi^*(z) \in B(G)$ *are extremal in* (2.13) *if and only if*

$$f^*(\zeta)[\omega(\zeta) - \phi^*(\zeta)]\,d\zeta = \lambda e^{i\delta}\frac{|f^*(\zeta)|}{\rho(\zeta)}\,ds \tag{4.1}$$

almost everywhere on Γ (δ *is real*; λ *is a common value for both extrema in* (2.13)), *and also*

$$\int_\Gamma \left|\frac{f^*(\zeta)}{\rho(\zeta)}\right| ds = 1. \tag{4.2}$$

3. *The extremal function* $\phi^*(z)$ *is always unique provided that an extremal function* $f^*(z)$ *exists. Otherwise it need not be unique. An extremal function* $f^*(z)$ *is unique in some cases* (*up to a factor* $e^{i\delta}$) *and in some cases is not.*

PROOF. The existence of an extremal function $\phi^*(z)$ has already been proved. If $\omega(\zeta)$ and $\rho(\zeta)$ are continuous, the existence of the extremal function $f^*(z) \in E^1_{\rho(\zeta)}$ can be proved in the same manner as the existence of $f^*(z) \in B^1_{\rho(\zeta)}$ in the previous section (we apply Lemma 4.1 here).

Let $f^*(z)$ and $\phi^*(z)$ be extremal functions in (2.13) (assuming that $f^*(z)$ exists). We have

$$\begin{aligned}\left|\int_\Gamma f^*(\zeta)\omega(\zeta)\,d\zeta\right| &= \left|\int_\Gamma f^*(\zeta)[\omega(\zeta) - \phi^*(\zeta)]\,d\zeta\right| \\ &\leqslant \int_\Gamma |f^*(\zeta)[\omega(\zeta) - \phi^*(\zeta)]|\,|d\zeta| \\ &\leqslant \operatorname{ess\,sup}\left[\rho(\zeta)|\omega(\zeta) - \phi^*(\zeta)|\right] \cdot \int_\Gamma \left|\frac{f^*(\zeta)}{\rho(\zeta)}\right| ds \leqslant \lambda.\end{aligned}$$

On the other hand, all inequalities in this chain must be equalities. This is only possible if

$$\rho(\zeta)|[\omega(\zeta) - \phi^*(\zeta)]| = \lambda \text{ a.e.}; \tag{4.3}$$

$$\arg\{f^*(\zeta)[\omega(\zeta) - \phi^*(\zeta)]\,d\zeta\} = \delta \text{ a.e.}; \tag{4.4}$$

$$\int_\Gamma \left|\frac{f^*(\zeta)}{\rho(\zeta)}\right| ds \leqslant 1.$$

From (4.3) and (4.4) it follows that (4.1) holds almost everywhere on Γ.

Conversely, let two functions $f^*(z) \in E^1_{\rho(\zeta)}$ and $\phi^*(z) \in B(G)$ satisfy (4.1) almost everywhere on Γ and

$$\int_\Gamma \left|\frac{f^*(\zeta)}{\rho(\zeta)}\right| ds = 1.$$

Then

$$\begin{aligned}\left|\int_\Gamma f^*(\zeta)\omega(\zeta)\,d\zeta\right| &= \left|\int_\Gamma f^*(\zeta)[\omega(\zeta) - \phi^*(\zeta)]\,d\zeta\right| \\ &= \lambda\int_\Gamma \frac{|f^*(\zeta)|}{\rho(\zeta)}\,ds = \lambda.\end{aligned}$$

On the other hand, for any function $f(z) \in E^1_{\rho(\zeta)}$ we have

$$\left|\int_\Gamma f(\zeta)\omega(\zeta)\,d\zeta\right| = \left|\int_\Gamma f(\zeta)[\omega - \phi^*(\zeta)]\,d\zeta\right| \leqslant \lambda.$$

Thus,

$$\lambda = \sup_{f \in E^1_{\rho(\zeta)}} \left|\int_\Gamma f(\zeta)\omega(\zeta)\,d\zeta\right|$$

and the supremum is attained for $f^*(z)$. Moreover, since

$$\lambda = \operatorname{ess\,sup} \rho(\zeta)|\omega(\zeta) - \phi^*(\zeta)|,$$

it follows from (2.13) that

$$\lambda = \inf_{\phi \in B(G)} \operatorname{ess\,sup} \rho(\zeta)|\omega(\zeta) - \phi(\zeta)|.$$

Assertion 2 of Theorem 4.2 is proved.

To begin the proof of assertion 3, let us show the uniqueness of $\phi^*(z)$ provided that $f^*(z)$ does exist. Indeed, taking a specific extremal function $f^*(z)$, we derive from (4.1) that

$$\arg[\omega(\zeta) - \phi^*(\zeta)] = \delta - \arg f^*(\zeta) - \arg d\zeta$$

almost everywhere on Γ. So, $\arg[\omega(\zeta) - \phi^*(\zeta)]$ is uniquely defined. Also, $|\omega(\zeta) - \phi^*(\zeta)| = \lambda/\rho(\zeta)$ almost everywhere. This shows that the difference $\omega(\zeta) - \phi^*(\zeta)$ is uniquely defined. Hence, $\phi^*(\zeta)$ is unique.

Let us make a few remarks concerning uniqueness of the extremal function $f^*(z)$. First of all, we note that any function $f(z) \in E_1$ such that

$$\int_\Gamma \left| \frac{f(\zeta)}{\rho(\zeta)} \right| ds = 1$$

is extremal for a problem of

$$\sup_{f \in E^1_{\rho(\zeta)}} \left| \int_\Gamma f(\zeta)\omega(\zeta)\, d\zeta \right|$$

with a bounded function $\omega(\zeta)$. Define $\omega(\zeta)$ by

$$\omega(\zeta) = [\rho(\zeta)]^{-1} \cdot e^{-i\arg[f(\zeta)\, d\zeta/ds]}. \tag{4.5}$$

Then, almost everywhere on Γ,

$$f(\zeta)\omega(\zeta)\, d\zeta = (|f(\zeta)|/\rho(\zeta))\, ds. \tag{4.6}$$

By virtue of assertion 2 of our theorem, this equality shows that $f(z)$ is extremal ($\phi^*(z) \equiv 0$).

Let Γ consist of analytic contours. Consider the functions $w_k'(z)$ used at the end of the previous section (in (3.3)). $w_k'(z)$ is analytic and single-valued in $\overline{G}$, and $\operatorname{Im}(iw_k'(\zeta)\, d\zeta) = 0$ on Γ. Moreover,

$$iw_k'(\zeta)\frac{d\zeta}{ds} < 0, \qquad \zeta \in \gamma_k \tag{4.7}$$

and

$$iw_k'(\zeta)\frac{d\zeta}{ds} > 0, \qquad \zeta \in \gamma_j,\ j \neq k. \tag{4.8}$$

In fact,

$$\begin{aligned} iw_k'(\zeta)\frac{d\zeta}{ds} &= i\frac{dw_k(\zeta)}{ds} = i\frac{\partial}{\partial s}[\omega(\zeta) + i\omega_k^*(\zeta)] \\ &= -\frac{\partial}{\partial s}\omega_k^*(\zeta) = \frac{\partial}{\partial n}\omega_k(\zeta), \end{aligned}$$

where n is the inward normal to Γ.

Recalling that $\omega_k(z)$ is a harmonic function defined on Γ by

$$\omega_k(\zeta) = 1, \quad \zeta \in \gamma_k; \qquad \omega_k(\zeta) = 0, \quad \zeta \in \Gamma \setminus \gamma_k,$$

we obtain, by the maximum principle, that

$$iw_k'(\zeta)\frac{d\zeta}{ds} \leqslant 0, \quad \zeta \in \gamma_k; \qquad iw_k'(\zeta)\frac{d\zeta}{ds} \geqslant 0, \quad \zeta \in \gamma_j, \ j \neq k.$$

Let us exclude the possibility of vanishing. The analytic function $w = w_k(z)$ gives a (multivalued) mapping of the domain onto a strip in the plane w: $0 \leqslant \operatorname{Re} w \leqslant 1$. Suppose that $w_k'(\zeta_0) = 0$ at a certain point $\zeta_0 \in \Gamma$. Then in a neighborhood of ζ_0 the mapping is at least two-sheeted. The image of that part of a sufficiently small neighborhood of ζ_0 which belongs to G cannot fit into the strip $0 \leqslant \operatorname{Re} w \leqslant 1$. Hence, the equality $w_k'(\zeta) = 0$ cannot occur. Thus, the inequalities (4.7) and (4.8) have been established.

Take two functions $w_k'(z)$ and $w_j'(z)$, $k \neq j$. Let $0 < m = \min_{\zeta \in \Gamma}|w_k'(\zeta)|$ and $M = \max_{\zeta \in \Gamma}|w_j'(z)|$. Fix two numbers $0 < \lambda_1 < \lambda_2 < 1$ and construct functions

$$f_{\lambda_i}(z) = C_i\left[w_k'(z) + \lambda_i(m/M)w_j'(z)\right], \qquad i = 1, 2. \tag{4.9}$$

The constants $C_i > 0$ are chosen in such a way that

$$\int_\Gamma \frac{|f_{\lambda_i}(\zeta)|}{\rho(\zeta)}\, ds = 1.$$

Construct the function $\omega(\zeta)$ according to (4.5) for $f_{\lambda_1}(z)$. For that $\omega(\zeta)$ the function $f_{\lambda_1}(z)$ is extremal. We have on Γ

$$f_{\lambda_1}(\zeta)\omega(\zeta)\, d\zeta = \frac{|f_{\lambda_1}(\zeta)|}{\rho(\zeta)}\, ds.$$

The function $f_{\lambda_2}(z)$ (for any λ_2, $0 < \lambda_2 < 1$) together with $f_{\lambda_1}(z)$ gives a fraction

$$\frac{f_{\lambda_1}(\zeta)}{f_{\lambda_2}(\zeta)} = \frac{C_1\left[w_k'(\zeta) + \lambda_1(m/M)w_j'(\zeta)\right]}{C_2\left[w_k'(\zeta) + \lambda_2(m/M)w_j'(\zeta)\right]} > 0$$

on Γ. Therefore,

$$f_{\lambda_1}(\zeta)\frac{f_{\lambda_2}(\zeta)}{f_{\lambda_1}(\zeta)}\omega(\zeta)\, d\zeta = \frac{|f_{\lambda_1}(\zeta)|}{\rho(\zeta)} \cdot \left|\frac{f_{\lambda_2}(\zeta)}{f_{\lambda_1}(\zeta)}\right| ds,$$

i.e.,

$$f_{\lambda_2}(\zeta)\omega(\zeta)\, d\zeta = \frac{|f_{\lambda_2}(\zeta)|}{\rho(\zeta)}\, ds.$$

This inequality shows that $f_{\lambda_2}(z)$ is also an extremal function. We have constructed an example where an extremal function for the problem of finding

$$\sup_{f \in E^1_{\rho(\zeta)}} \left|\int_\Gamma f(\zeta)\omega(\zeta)\, d\zeta\right|$$

is not unique.

For some problems the extremal function $f^*(z) \in E^1_{\rho(\zeta)}$ is unique up to a factor $e^{i\delta}$. Indeed, assume that some $f^*(z)$ such that $|f^*(z)| \geqslant \tau > 0$ in $\bar{G}$ is an extremal function for a certain $\omega(\zeta)$. (For a given $f^*(z)$ with such properties a function $\omega(\zeta)$, for which $f^*(z)$ is extremal, can be constructed according to (4.5).) Uniqueness (up to a factor $e^{i\delta}$) of $f^*(z)$ can easily be derived from (4.1).

To finish the proof of assertion 1 of Theorem 4.2, let us show that there exist bounded functions $\omega(\zeta)$ such that $\sup_{f \in E^1_{\rho(\zeta)}} |\int_\Gamma f(\zeta)\omega(\zeta)\, d\zeta|$ is not attained for any $f(z)$. Let G be simply connected and, as usual, let Γ be its boundary. Consider the class E^1_1 ($\rho(\zeta) \equiv 1$). Take an arc $\gamma \subset \Gamma$ and set

$$\omega(\zeta) = \begin{cases} 1, & \zeta \in \gamma, \\ 0, & \zeta \in \Gamma \setminus \gamma. \end{cases}$$

Let α and β be the endpoints of γ. Set

$$\lambda = \sup_{f \in E^1_1} \left| \int_\Gamma f(\zeta)\omega(\zeta)\, d\zeta \right| = \sup_{f \in E^1_1} \left| \int_\gamma f(\zeta)\, d\zeta \right|.$$

According to (2.13)

$$\lambda = \inf_{\phi \in B(G)} \operatorname{ess\,sup} |\omega(\zeta) - \phi(\zeta)| \leqslant \operatorname{ess\,sup} |\omega(\zeta) - \tfrac{1}{2}| = \tfrac{1}{2}.$$

Fix an arbitrary $\varepsilon > 0$. Take a segment of length $1/2 - \varepsilon$ in the w-plane and surround it by an analytic contour L of length 1.

Consider $w = \psi(z)$ mapping G onto a domain D in the interior of L such that α and β are mapped into the points A and B of L so that the distance between A and B is equal to the diameter of L. The following inequality is obvious:

$$|\psi(\alpha) - \psi(\beta)| = |A - B| > \tfrac{1}{2} - \varepsilon.$$

The function $\psi'(z)$ belongs to E_1 in G, since $\psi(z)$ is continuous in $\bar{G}$ and absolutely continuous on Γ. Also, $\int_\Gamma |\psi'(\zeta)||d\zeta| = 1$, and so $\psi' \in E^1_1$. Furthermore,

$$\lambda \geqslant \left| \int_\gamma \psi'(\zeta)\, d\zeta \right| = |\psi(\alpha) - \psi(\beta)| > \tfrac{1}{2} - \varepsilon.$$

Thus $\lambda = \frac{1}{2}$.

Assume that there exists a function $f^*(z)$ such that

$$\left| \int_\Gamma f^*(\zeta)\omega(\zeta)\, d\zeta \right| = \tfrac{1}{2}.$$

Then, almost everywhere on Γ,

$$f^*(\zeta)[\omega(\zeta) - \phi^*(\zeta)]\, d\zeta = \tfrac{1}{2}|f^*(\zeta)|\, ds.$$

But $\phi^*(z) \equiv \frac{1}{2}$. Therefore $\pm f^*(\zeta)\, d\zeta = |f^*(\zeta)|\, ds$, i.e., $f^*(\zeta)\, d\zeta$ is real-valued on Γ. Arguing as in §3 we conclude that $f^*(z) \equiv 0$. This is the desired contradiction.

Introducing slight changes in the above example, we can construct an example of a problem for which the extremal function $\phi^*(z) \in B(G)$ is not unique. Let G

be the unit disk. Take $\omega(\zeta) = \omega(e^{i\theta})$ defined by

$$\omega(e^{i\theta}) = \begin{cases} 1, & 0 \leqslant \theta \leqslant \pi/2; \\ -1, & \pi/2 \leqslant \theta < \pi; \\ 0, & \pi \leqslant \theta < 2\pi, \end{cases} \tag{4.10}$$

and consider the problem of

$$\lambda = \sup_{f \in E_1^1} \left| \int_0^{2\pi} f(\zeta)\omega(\zeta)\, d\zeta \right|, \qquad \zeta = e^{i\theta}. \tag{4.11}$$

It is clear that $\lambda \leqslant 1 = \operatorname{ess\,sup}|\omega(e^{i\theta})|$. On the other hand, take the segment $[-\frac{1}{2} + \varepsilon, \frac{1}{2} - \varepsilon]$ in the w-plane ($\varepsilon > 0$ is arbitrarily small) and surround it by an analytic contour L of length 2. Let $w = \psi(z)$ map G onto the domain D bounded by L. Then $\frac{1}{2}\psi'(z) \in E_1^1(G)$. Let the points 1, i and -1 of the unit circle be mapped by $\omega = \psi(z)$ into the points A, B and C respectively, chosen so that $|A - B| > 1 - \varepsilon$, $|C - B| > 1 - \varepsilon$ and the length of the arc AC is arbitrarily small. It is obvious that we can always achieve this. Then

$$\left| \int_C^{2\pi} \frac{\psi'(\zeta)}{2}\omega(\zeta)\, d\zeta \right| = \frac{1}{2}|(B - A) - (C - B)| > 1 - 2\varepsilon.$$

The last inequality shows that $\lambda = 1$.

It has been already mentioned that $1 = \operatorname{ess\,sup}|\omega(\zeta)|$ and, therefore, the function $\phi^*(z) \equiv 0$ is extremal for the right-hand side of the duality relation (2.13). Let us construct another extremal function $\phi_1^*(z)$. For that purpose, let us map the disk $|z| < 1$ onto a semidisk with the unit radius in a lower half-plane w. Let $\phi_1^*(z)$ be a mapping function normalized by the following conditions:

$$\psi_1^*(1) = 1, \quad \phi_1^*(i) - 0, \quad \phi_1^*(-1) = -1$$

Then it is clear that $\phi_1^*(z) \in B$ and the following relations hold:

$$\begin{aligned} 0 \leqslant \phi_1^*(e^{i\theta}) &\leqslant 1, & 0 \leqslant \theta &< \pi/2; \\ 1 \leqslant \zeta_1^*(e^{i\theta}) &< 0, & \pi/2 \leqslant \theta &< \pi; \\ |\zeta_1^*(e^{i\theta})| &= 1, & \pi \leqslant \theta &< 2\pi. \end{aligned}$$

From this we see that $\operatorname{ess\,sup}|\omega(\zeta) - \phi_1^*(\zeta)| = 1$ and, therefore, $\phi_1^*(z)$ is also an extremal function for (2.13).

THEOREM 4.3. *Let functions $\omega(\zeta) \in L_p(G)$ and $\rho(\zeta)$ such that $0 < m \leqslant \rho(\zeta) \leqslant M < +\infty$ be given on the boundary of a domain G satisfying the Smirnov condition (we exclude the trivial case $\omega(\zeta) \in E_p(G)$). Then the followng assertions are true:*

1. *There exist functions $f^*(z) \in E^1_{\rho(\zeta),q}$ and $\phi^*(z) \in E_p(G)$ extremal for* (2.20).
2. *The functions $f^*(z) \in E^1_{\rho(\zeta),q}$ and $\phi^*(z) \in E_p(G)$ are extremal in* (2.20), *if and only if*

$$f^*(\zeta)[\omega(\zeta) - \phi^*(\zeta)]\, d\zeta = \frac{e^{i\delta}}{\lambda^{p-1}}\left[\rho(\zeta)|\omega(\zeta) - \phi^*(\zeta)|\right]^p ds \tag{4.12}$$

almost everywhere on Γ, *or, in a different form*,

$$f^*(\zeta)[\omega(\zeta)-\phi^*(\zeta)]\,d\zeta=\lambda e^{i\delta}\left(\frac{|f^*(\zeta)|}{\rho(\zeta)}\right)^q ds. \tag{4.13}$$

3. *The extremal functions* $\phi^*(z)$ *and* $f^*(z)$ *are unique (the latter up to a constant factor* $e^{i\delta}$*).*

The proof is the same as above.

§5. Local properties of extremal functions

THEOREM 5.1. *Assume that under the hypothesis of Theorem* 3.2 *the function* $\omega(\zeta)$ *coincides with boundary values of the analytic function* $\omega(z)$ *a.e. on some arc* $\gamma\subset\Gamma$. *Assume also that* $\omega(z)\in E_1(\tilde G)$ *in some domain* $\tilde G\subset G$ *such that* $\gamma\subset\partial\tilde G$. *If* $\tilde{\tilde G}\subset\tilde G$ *is an arbitrary domain sufficiently closely attached to* γ *along a certain arc* γ' *interior to* γ, *then*

$$\frac{1}{f^*(z)}\in D(\tilde{\tilde G}),\qquad \frac{\alpha'(z)}{\omega(z)-\phi^*(z)}\in D(\tilde{\tilde G}). \tag{5.1}$$

Here $f^*(z)$ *and* $\phi^*(z)$ *are extremal functions and* $w=\alpha(z)$ *is a conformal mapping of* G *onto a canonical circular domain.*

Let us start with the following simple lemma.

LEMMA 5.2. *If* $f(z)\in E_1(G)$ *and* γ *is an arc contained in the boundary* Γ *of the domain* G, *then for any* $\varepsilon>0$ *there exist an arc* $\gamma'\supset\gamma$ *and domain* $G^*\subset G$, $\gamma'\subset\partial G^*$, *such that* $\operatorname{meas}(\gamma'-\gamma)<\varepsilon$ *and* $f(z)\in E_1(G)$.

PROOF. To show that this is true, let us take any $\gamma'\supset\gamma$ satisfying in addition to the condition $\operatorname{meas}(\gamma'-\gamma)<\varepsilon$ only the requirement that $f(z)$ has finite angular boundary values at the endpoints ζ_1 and ζ_2 of γ'. For G^* we can now take the domain bounded by γ' and γ'', where γ'' is a rectifiable curve in G with endpoints ζ_1 and ζ_2, and nontangential to Γ at ζ_1 and ζ_2. For the contours Γ_j^* we take the arcs γ'' and $\Gamma_j\cap G^*$, where the Γ_j are the contours which determine the inclusion of $f(z)$ in $E_1(G)$. It is immediately seen that the integrals $\int_{\Gamma_j^*}|f(z)||dz|$ are uniformly bounded and, therefore, $f(z)\in E_1(G^*)$. Let us proceed to the proof of Theorem 5.1. Clearly, $f^*(z)\phi^*(z)\in E_1(G)$. Also, by hypothesis $\omega(z)\in E_1(\tilde G)$, and hence $f^*(z)\omega(z)\in E_1(\tilde G)$. By Lemma 5.2 and the argument in its proof, it suffices to consider an arc $\gamma'\subset\gamma$ to which we can attach a domain $\tilde G_1$ such that

$$f^*(z)\phi^*(z)\in E_1(\tilde G_1)\quad\text{and}\quad f^*(z)\omega(z)\in E_1(\tilde G_1)$$

simultaneously. Hence, $f^*(z)[\omega(z)-\phi^*(z)]\in E_1(\tilde G_1)$. We now use the fact that the equality (3.1) holds almost everywhere on Γ:

$$f^*(\zeta)[\omega(\zeta)-\phi^*(\zeta)]\,d\zeta=\rho(\zeta)|\omega(\zeta)-\phi^*(\zeta)|\,ds \tag{5.2}$$

(for simplicity we have put $e^{i\delta}=1$ in (3.1)). In particular, we have

$$f^*(\zeta)[\omega(\zeta)-\phi^*(\zeta)]\,d\zeta\geqslant 0 \tag{5.3}$$

on γ'. Let us map $\tilde{G}_1$ conformally onto the disk K: $|w| < 1$ by means of the functions $z = \beta_1(w)$, $w = \alpha_1(z)$. The function

$$R(w) = f^*(\rho_1(w))[\omega(\beta_1(w)) - \phi^*(\beta_1(w))]\beta_1'(w) \tag{5.4}$$

belongs to the class $H_1(K)$ in K, since $f^*(z)[\omega(z) - \phi^*(z)] \in E_1(\tilde{G}_1)$. Also, $R(w)\,dw \geqslant 0$ on the arc Δ of a unit circle, which is an image of γ'. Therefore $R(w)$ can be analytically continued across the open arc Δ. (We have already mentioned in §3 that the properties of the class E_1 allow us to extend immediately the classical proof of the symmetry principle to this class. But $H_1 = E_1$ in K.) Let us return to the domain $\tilde{G}_1$. Since we can analytically continue $R(w)$ across Δ, we conclude that for any arc $\gamma'' \subset \gamma'$ there exists a domain $\tilde{\tilde{G}} \subset \tilde{G}_1$ attached to it such that $R(\alpha_1(z))$ is analytic, continuous up to the boundary and not vanishing in it. Therefore, $[R(\alpha_1(z))]^{-1}$ belongs to the class $D(\tilde{\tilde{G}})$ (see the Appendix); possible existence of a few zeros of $R(\alpha_1(z))$ on the boundary cannot affect this inclusion. From (5.4) we have

$$\frac{1}{f^*(z)} = \frac{\omega(z) - \phi^*(z)}{R(\alpha_1(z))\alpha_1'(z)}. \tag{5.5}$$

The function $\omega(z) - \phi^*(z)$ belongs to $E_1(\tilde{G})$, and by Lemma 5.2, we can assume that $\omega(z) - \phi^*(z) \in E_1(\tilde{G}_1)$. Then, according to Theorem 4.1,

$$\frac{\omega(z) - \phi^*(z)}{\alpha_1'(z)} \in H_1(\tilde{G}_1) \subset D(\tilde{G}_1) \subset D(\tilde{\tilde{G}}).$$

Also, $[R(\alpha_1(z))]^{-1} \subset D(\tilde{\tilde{G}})$, and therefore $1/f^*(z) \in D(\tilde{\tilde{G}})$. From (5.5) we obtain

$$\frac{\alpha_1'(z)}{\omega(z) - \phi^*(z)} = \frac{f^*(z)}{R(\alpha_1(z))} \in D(\tilde{\tilde{G}}). \tag{5.6}$$

But instead of the derivative $\alpha_1'(z)$ of the conformal mapping of the subdomain $\tilde{\tilde{G}}$ onto the unit disk, we can take the derivative of the conformal mapping of G onto a canonical circular domain. It might happen that we will have to decrease $\tilde{\tilde{G}}$ slightly by taking the domain $\tilde{\tilde{G}}^*$ such that its boundary contains an arc $\gamma'' \subset \gamma'$ whose length differs from the length of γ' by an arbitrarily small number. This proves that

$$\frac{\alpha'(z)}{\omega(z) - \phi^*(z)} \in D(\tilde{\tilde{G}}^*)$$

and the theorem is proved.

REMARK 5.3. Relations (5.1) mean that extremal functions approach the boundary "smoothly". If we do not make the assumptions indicated in Theorem 5.1, then, for example, $1/f^*(z)$ need not belong to $D(\tilde{\tilde{G}})$. Indeed, any analytic function $f_0(z)$ in the class B_ρ^1 such that $|f_0(\zeta)| = \rho(\zeta)$ a.e. on Γ is an extremal function in a certain problem of the type under consideration with a specially chosen $\omega(\zeta)$; cf. (4.5). But $f_0(z)$ may be such that $1/f_0(z)$ does not belong to the class D in any of the subdomains attached to a certain arc of the boundary Γ.

For simplicity, we consider the case where G is the unit disk. Construct the analytic function $f_0(z)$ as follows:

$$f_0(z) = \exp\frac{1}{2\pi}\int_0^{2\pi}\frac{e^{i\theta}+z}{e^{i\theta}-z}[\ln\rho(\theta)\,d\theta + d\mu(\theta)],$$

where $d\mu(\theta) \leqslant 0$ is a singular measure with set of points of decrease everywhere dense on $[0, 2\pi]$. Then $f_0(z)$ is a bounded function in the unit disk, and $|f_0(\zeta)| = \rho(\zeta)$ a.e. on Γ. But if $1/f_0(z)$ belonged to the class D in some subdomain $\tilde{\tilde{G}}$ attached to the unit circle Γ along the arc $\gamma \subset \Gamma$, then we would have $\mu(\theta) \equiv \text{const}$ on γ, which is not true. In the same way if we do not impose additional conditions on $\omega(\zeta)$, then, generally speaking,

$$\frac{\alpha'(z)}{\omega(z) - \phi^*(z)} \notin D(\tilde{\tilde{G}}).$$

REMARK 5.4. From the inclusion

$$\frac{\alpha'(z)}{\omega(z) - \phi^*(z)} \in D(\tilde{\tilde{G}})$$

it immediately follows that

$$\frac{1}{\omega(z) - \phi^*(z)} \in D(\tilde{\tilde{G}}),$$

since $1/\alpha'(z) = \beta'(\alpha(w)) \subset H_1(G) \subset D(G) \subset D(\tilde{\tilde{G}})$ (see the Appendix). Here, $z = \beta(w)$ is the inverse function of $\alpha(z)$. However, the assertion

$$\frac{\alpha'(z)}{\omega(z) - \phi^*(z)} \in D(\tilde{\tilde{G}})$$

is stronger, because, in general, $\alpha'(z)$ does not belong to $D(\tilde{\tilde{G}})$.

Using an argument similar to the proof of Theorem 5.1, we can prove the following theorems.

THEOREM 5.5. *Under the hypothesis of Theorem* 4.2, *let the function* $\omega(\zeta)$ *coincide almost everywhere on an arc* $\gamma \subset \Gamma$ *with the boundary values of a function* $\omega(z)$ *which is bounded and analytic in a domain* $\tilde{G} \subset G$ *attached to* γ. *Also, assume that an extremal function* $f^*(z)$ *exists. Then*

$$\frac{\alpha'(z)}{f^*(z)} \in D(\tilde{\tilde{G}}), \qquad \frac{1}{\omega(z) - \phi^*(z)} \in D(\tilde{\tilde{G}}), \tag{5.7}$$

where $\tilde{\tilde{G}}$ *and* $\alpha(z)$ *are the same as in Theorem* 5.1.

THEOREM 5.6. *Under the hypothesis of Theorem* 4.3, *let the function* $\omega(\zeta)$ *coincide almost everywhere on some arc* $\gamma \subset \Gamma$ *with the boundary values of some analytic function* $\omega(z) \in E_p(\tilde{G})$, *where* $\tilde{G}$ *is a subdomain of* G *attached to* γ. *Then*

$$\frac{\sqrt[q]{\alpha'(z)}}{f^*(z)} \in D(\tilde{\tilde{G}}), \qquad \frac{\omega(z) - \phi^*(z)}{\sqrt[p]{\alpha'(z)}} \in D(\tilde{\tilde{G}}), \tag{5.8}$$

where $\tilde{\tilde{G}}$ *and* $\omega(z)$ *are the same as in Theorem* 5.1.

THEOREM 5.7. *If in Theorems* 5.1, 5.5, *and* 5.6 γ *coincides with the whole boundary contour* $\gamma_i \subset \Gamma$, *then the assertions of those theorems hold for a domain* $\tilde{\tilde{G}}$ *attached to the whole of* γ_i.

PROOF. Analyzing the argument given above, it is easy to see that in Lemma 5.2 for the case $\gamma = \gamma_i$ we can take for G^* a domain attached from G to the whole of γ_i. Also, in this case the ratio of the derivatives of the conformal mappings of the domains attached to γ_i is bounded near the whole curve γ_i.

THEOREM 5.8. *If the hypothesis of Theorem* 5.1 *is satisfied, then the extremal function* $f^*(z)$ *in* $\tilde{\tilde{G}}$ *can be represented in the following form*:

$$f^*(z) = \exp\left[\frac{1}{2\pi}\int_\gamma \ln\rho(\zeta)\frac{\partial P(z,\zeta)}{\partial n}\,ds\right]\cdot F(z), \tag{5.9}$$

where $P(\zeta, z)$ *is the complex Green function of* G, $F(z)$ *is analytic in* $\tilde{\tilde{G}}$, *continuous up to* γ *and* $|F(\zeta)| = 1$, $\zeta \in \gamma$.

PROOF. $f^*(z) \neq 0$ in $\tilde{\tilde{G}}$ and, therefore, $\ln|f^*(z)|$ is a harmonic function. Since $f^*(z)$ and $1/f^*(z)$ belong to $D(\tilde{\tilde{G}})$, the families of integrals

$$\int_e \ln^+|f^*(z)|\frac{\partial g^j}{\partial n}\,ds, \qquad \int_e \ln^+\left|\frac{1}{f^*(z)}\right|\frac{\partial g^j}{\partial n}\,ds \tag{5.10}$$

are absolutely equicontinuous with respect to the harmonic measure (see the Appendix). Therefore, the family

$$\int_e \ln|f^*(z)|\frac{\partial g^j}{\partial n}\,ds \tag{5.11}$$

is also absolutely equicontinuous. Here, as usual, we have constructed a sequence of domains $\{G^j\}$ exhausting $\tilde{\tilde{G}}$, and the functions g^j are the Green functions of G^j with the pole at a fixed point. The sets e are taken on the boundaries of the G^j. From the absolute equicontinuity of the family (5.11) it follows that the harmonic function $\ln|f^*(z)|$ is representable by the Green formula in $\tilde{\tilde{G}}$ (see the Appendix). Consider the harmonic function

$$u(z) = \frac{1}{2\pi}\int_\gamma \ln\rho(\zeta)\frac{\partial g(\zeta,z)}{\partial n}\,ds \tag{5.12}$$

in G, where $g(\zeta, z)$ is the Green function of G. Being representable by the Green formula in G, the function $u(z)$ is also representable by this formula in each subdomain of G; in particular, in $\tilde{\tilde{G}}$. Hence, the difference

$$U(z) = \ln|f^*(z)| - u(z) \tag{5.13}$$

is representable by the Green formula in G. Also, on γ we have

$$U(\zeta) = \ln\rho(\zeta) - u(\zeta) = \ln\rho(\zeta) - \ln\rho(\zeta) = 0. \tag{5.14}$$

Construct an analytic function $\Phi(z) = U(z) + iV(z)$ and set $F(z) = e^{\Phi(z)}$. Then $F(z)$ is continuous in $\tilde{\tilde{G}}$ up to γ. This will be proved if we show that $\Phi(z)$ is continuous up to γ. Let us map the unit $|w| < 1$ onto $\tilde{\tilde{G}}$. By means of the function $z = \alpha_1(w)$ we obtain the function

$$\Phi(\alpha_1(w)) = U(\alpha_1(w)) + iV(\alpha_1(w)) \tag{5.15}$$

in the disk, and $U(\alpha_1(w))$ is representable by the Poisson integral. According to (5.14) this function vanishes on the arc Δ, the image of γ. But then $U(\alpha_1(w))$ can be extended as a harmonic function across Δ, and $\Phi(\alpha_1(w))$ is continuous up to Δ. Hence, $\Phi(z)$ and $F(z)$ are continuous up to γ, and $|F(\zeta)| = 1$ on γ. From (5.12) it follows that the function $v(z)$ conjugate to $u(z)$ can be represented in the form

$$v(z) = \frac{1}{2\pi}\int_\gamma \ln\rho(\zeta)\frac{\partial\tilde{g}(z,\zeta)}{\partial n}\,ds. \tag{5.16}$$

Here $\tilde{g}(\zeta, z)$ is conjugate to $g(\zeta, z)$ with respect to ζ (and, therefore, $\tilde{g}(z,\zeta)$ is conjugate to $g(z,\zeta) = g(\zeta, z)$ with respect to z). Hence,

$$u(z) + iv(z) = \frac{1}{2\pi}\int_\gamma \ln\rho(\zeta)\frac{\partial P(z,\zeta)}{\partial n}\,ds, \tag{5.17}$$

where $P(\zeta, z) = g(\zeta, z) + i\tilde{g}(\zeta, z)$ is the complex Green function in G. From all this, and, in particular, from (5.17), the assertion of the theorem follows.

REMARK 5.9. If γ coincides with the whole boundary contour γ_i, then from Theorem 5.7 we conclude that (5.9) holds in a domain $\tilde{\tilde{G}} \subset G$ attached to the whole of γ_i. However, $F(z)$ now need not be single-valued in $\tilde{\tilde{G}}$, in general, although its modulus is always single-valued.

REMARK 5.10. We could give similar representations for extremal functions, and under the hypotheses of Theorems 5.5 and 5.6 these representations will show that extremal functions do not contain factors with singular measures.

The derivation of the corresponding formulas is left to the reader.

THEOREM 5.11. *If under the hypothesis of Theorem* 5.1, 5.5, *or* 5.6 γ *is an analytic arc, then the product* $f^*(z)[\omega(z) - \phi^*(z)]$ *can be analytically extended across* γ.

A proof follows immediately from (5.3) and similar inequalities for Theorems 5.5 and 5.6. (Actually, we have already used this fact repeatedly.)

THEOREM 5.12. *Under the hypothesis of Theorem* 5.1, *the modulus of the extremal function* $f^*(z)$ *is continuous up to* γ.

Indeed, $\rho(\zeta)$ is a continuous function. Then the Green integral (5.12) is continuous up to γ, as follows, for instance, from the proof of Theorem 5.8. However, this is a well-known property of Green integrals.

THEOREM 5.13. *Under the hypothesis of Theorem* 5.1, *let the arc* γ *be analytic and let* $\ln \rho(\zeta)$ *be an analytic function of the arclength* s *on* γ. *Then the extremal function* $f^*(z)$ *and the extremal difference* $\omega(z) - \phi^*(z)$ *can be analytically continued across* γ.

PROOF. The harmonic function $u(z)$ in (5.12) is continuous up to γ, and its value on γ is an analytic function of the arclength s. Therefore, $u(z)$ can be harmonically continued across γ, and $u + iv$ in (5.7) can be analytically continued across γ. The function $F(z)$ is continuous up to γ, and $|F(\zeta)| = 1$. From this it follows that $F(z)$ can also be analytically continued across γ. Hence, $f^*(z)$ can be analytically continued across γ. Taking into account that the product $f^*(z)[\omega(z) - \phi^*(z)]$ can also be continued across γ, we see that the same conclusion holds for $\omega(z) - \phi^*(z)$.

REMARK 5.14. Similar facts can also be established for extremal functions in other classes. However, in those classes $f^*(z)$ and $\omega - \phi^*$ may have branch points on γ. The fact is that in analogs of (5.9), $\ln|f^*(\zeta)[\omega(\zeta) - \phi^*(\zeta)]|$ appears together with $\ln \rho(\zeta)$.

§6. Representations of extremal functions in the whole domain

LEMMA 6.1. *Let there be given a function* $u(z)$ *harmonic in a finitely connected domain* G *with a Jordan boundary* $\Gamma = \bigcup_1^n \gamma_j$. *If* $u(z)$ *is representable by Green's formula in certain annular domains* $G^i \subset G$, $i = 1, \dots, n$, *attached to* γ_i *inside of* G, *then* $u(z)$ *is representable by Green's formula in the whole domain* G.

THEOREM 6.2. *Under the hypothesis of Theorem* 5.1, *let* $\omega(\zeta)$ *coincide a.e. on every boundary contour* γ_i, $i = 1, \dots, n$, *with boundary values of the function* $\omega_i(\zeta)$ *which is analytic and belongs to* $E_1(\tilde{G}_i)$ *in a domain* $\tilde{G}_i$ *attached to* γ_i. *Then the extremal function* $f^*(z)$ *has finitely many zeros* $\alpha_1, \dots, \alpha_k$ *in the domain, and*

$$f^*(z) = e^{i\delta} \exp\left[\frac{1}{2\pi}\int_\Gamma \ln \rho(\zeta) \frac{\partial P(z,\zeta)}{dn}\, ds - \sum_{j=1}^{k} P(z, \alpha_j)\right], \tag{6.1}$$

where δ *is a real constant.*

PROOF. According to Theorem 5.1, the function $1/f^*(z)$ is analytic near each of the boundary contours γ_j, $j = 1, \dots, n$. Therefore, $f^*(z)$ has finitely many zeros in G. Denote these zeros by $\alpha_1, \dots, \alpha_k$. The function

$$U(z) = \ln|f^*(z)| + \sum_1^k g(z, \alpha_j) \tag{6.2}$$

is harmonic in G. From Theorems 5.7 and 5.8 it follows that $\ln|f^*(z)|$ is representable by Green's formula in sufficiently narrow annuli $\tilde{\tilde{G}}^i$ attached to the contours γ_i, $i = 1, \dots, n$, respectively. The function $\sum_1^k g(z, \alpha_j)$ is continuous in those domains up to their boundaries. Therefore, $U(z)$ is representable by Green's formula in each annulus $\tilde{\tilde{G}}_i$, $i = 1, \dots, n$. Then, by Lemma 6.1, $U(z)$ is representable by Green's formula in G. Since $\ln|f^*(\zeta)| = \ln \rho(\zeta) = U(\zeta)$ on Γ, we have

$$U(z)=\frac{1}{2\pi}\int_\Gamma \ln\rho(\zeta)\frac{\partial g(\zeta,z)}{\partial n}\,ds. \tag{6.3}$$

It is easy to see that (6.3) is equivalent to (6.1).

(6.1) reduces our extremal problem to a simpler one with finitely many variables $\alpha_1,\dots,\alpha_k$; however, we have not yet estimated the number of variables.

It is also possible (cf. Remark 5.14) to give representations of extremal functions in other classes, assuming that $\omega(\zeta)$ is analytic in strips attached to all boundary contours. Those representations would show that extremal functions in problems with such $\omega(\zeta)$ do not have factors with singular measure in their structure. However, those formulas are going to be less effective than (6.1), because instead of the factor

$$\exp\left[\frac{1}{2\pi}\int_\Gamma \ln\rho(\zeta)\frac{\partial P(z,\zeta)}{\partial n}\,ds\right],$$

determined by the function $\rho(\zeta)$ defining the class, a factor of the type

$$\exp\left[\frac{1}{2\pi}\int_\Gamma \ln|f(\zeta)|\frac{\partial P(z,\zeta)}{\partial n}\,ds\right]$$

will appear in those classes, where $f(z)$ is precisely the function we are trying to represent.

In Theorem 6.2, the function $\omega(\zeta)$ is not necessarily a boundary value of the same meromorphic function in G with finitely many poles. But in some of the most important extremal problems we will be dealing precisely with such $\omega(\zeta)$. For example, this is the case in the problems of $\sup|f(\beta)|$, $\sup|f'(\beta)|$, $\sup|\sum_1^m\lambda_j f^{(\nu_j)}(\beta_j)|$, etc. In such cases we will give an estimate of the number of zeros of $f^*(z)$ and, moreover, we will obtain representations of extremal functions not only in the class B_ρ^1 but also in other classes. The tools for our representations will be Green's and Neumann's functions.

Let G be an n-connected domain whose boundary Γ consists of analytic contours. Let $\zeta\in G$ be a given point. The Neumann function $N(z,\zeta)$ of G is defined by

$$N(z,\zeta)=\ln(1/|z-\zeta|)+u(z), \tag{6.4}$$

where $u(z)$ is harmonic in $\overline{G}$ and

$$\partial N/\partial n=\text{const} \tag{6.5}$$

on Γ. ($\partial/\partial n$ denotes differentiation in the inward normal direction.) For $N(z,\zeta)$ to be uniquely defined, we have to add a normalizing condition to (6.5), e.g.

$$\int_\Gamma N(z,\zeta)\,ds_z=0. \tag{6.6}$$

From the Green formula it is easily deduced that the constant in (6.5) is equal to $2\pi/l$, where l is the length of Γ. It is not hard to prove (again by Green's formula) that $N(z,\zeta)=N(\zeta,z)$. Let $\tilde N(z,\zeta)$ be the conjugate function of $N(z,\zeta)$

with respect to the variable z. Introduce the following functions:

$$\begin{aligned} q(z,\zeta) &= N(z,\zeta) + i\tilde{N}(z,\zeta); \\ N(z,\zeta_1,\zeta_2) &= N(z,\zeta_1) - N(z,\zeta_2); \\ \tilde{N}(z,\zeta_1,\zeta_2) &= \tilde{N}(z,\zeta_1) - \tilde{N}(z,\zeta_2); \\ q(z,\zeta_1,\zeta_2) &= q(z,\zeta_1) - q(z,\zeta_2); \\ \psi(z,\zeta_1,\zeta_2) &= \exp[q(z,\zeta_1,\zeta_2)]. \end{aligned} \tag{6.7}$$

The function $\tilde{N}(z,\zeta_1,\zeta_2)$ does not have periods in traversing the boundary contours since on Γ we have

$$\frac{\partial \tilde{N}(z,\zeta_1,\zeta_2)}{\partial s} = -\frac{\partial N(z,\zeta_1,\zeta_2)}{\partial n} = \frac{2\pi}{l} - \frac{2\pi}{l} = 0. \tag{6.8}$$

Moreover, from (6.8) it follows that $\tilde{N}(z,\zeta_1,\zeta_2)$ is constant on each boundary contour. The function $q(z,\zeta_1,\zeta_2)$ has the following decomposition:

$$q(z,\zeta_1,\zeta_2) = \ln\frac{z-\zeta_2}{z-\zeta_1} + F(z), \tag{6.9}$$

where $F(z)$ is analytic and single-valued in $\overline{G}$, and $\operatorname{Im} q(z,\zeta_1,\zeta_2) = \text{const}$ when z runs over a boundary contour. The function $\psi(z,\zeta_1,\zeta_2)$ has a simple pole at ζ_1, is analytic and single-valued at the remaining points of $\overline{G}$, has a simple zero at ζ_2, and, as is easily seen in accordance with the argument principle, gives a schlicht conformal mapping of G onto the plane with radial cuts.

When the pole ζ of the Green function $g(z,\zeta)$ moves to the boundary, the function becomes identically zero. This is not so for the Neumann function $N(z,\zeta_1,\zeta_2)$. As ζ_1 or ζ_2 moves to Γ, the main part in the decomposition of N takes the form

$$\ln\frac{|z-\zeta_2|}{|z-\zeta_1|^2}, \qquad \ln\frac{|z-\zeta_2|^2}{|z-\zeta_1|}, \qquad \ln\left|\frac{z-\zeta_2}{z-\zeta_1}\right|^2 \tag{6.10}$$

instead of $\ln|(z-\zeta_2)/(z-\zeta_1)|$, if, respectively, ζ_1, ζ_2, or both of those points reaches the boundary.

At the same time (6.8) still holds. The function $\psi(z,\zeta_1,\zeta_2)$ has a double pole at ζ_1 if $\zeta_1 \in \Gamma$ and a double zero at ζ_2 if $\zeta_2 \in \Gamma$. Again, $\arg\psi = \tilde{N}(z,\zeta_1,\zeta_2) = \text{const}$ on each boundary component. The reason for the changes mentioned above in the construction of N and ψ when poles and zeros move to the boundary becomes especially clear if we consider our functions on the Schottky double (see [75], Chapter 2, §2) of G. Then the function ψ, for example, in addition to having a zero at ζ_1, will have a zero at the conjugate point ζ_1^* of the double by the symmetry principle. When $\zeta_1 \to \zeta_1^0 \in \Gamma$, we have $\zeta_1^* \to \zeta_1^0$, and we get a double zero on the boundary of G.

THEOREM 6.3. *Let Γ, the boundary of G, consist of analytic contours. Suppose that the function $\omega(\zeta)$ is the boundary value on Γ of a meromorphic (in G) function $\omega(z)$ with poles $\beta_1, \dots, \beta_m$ (counted with multiplicities). Concerning $\omega(z)$ as z approaches Γ, the same behavior as in one of the Theorems* 5.1, 5.5 *or* 5.6 *is assumed, with the arc γ replaced now by the whole boundary Γ. Then under the hypotheses of all these theorems the product*

$$R(z) = f^*(z)[\omega(z) - \phi^*(z)] \tag{6.11}$$

is meromorphic in $\overline{G}$ with poles only at $\beta_1, \dots, \beta_m$ and has $m + n - 2$ zeros $\alpha_1, \dots, \alpha_{m+n-2}$ in $\overline{G}$. (Zeros on Γ have even order and are counted with half multiplicities.) If $\alpha = \alpha_{m+n-1}$ is an arbitrary point in G, and $\beta_{m+1}, \dots, \beta_{m+n-1}$ are the critical points of the Green function $g(z, \alpha)$ (i.e., zeros of $P'(z, \alpha)$), then the following formulas hold:

$$f^*(z)[\omega(z) - \phi^*(z)] = Cp'(z, \alpha) \prod_1^{m+n-1} \psi(z, \beta_j, \alpha_j); \tag{6.12}$$

$$\begin{aligned} f^*(z) = C_1 \exp&\left[\frac{1}{2\pi}\int_\Gamma \ln\rho(\zeta)\frac{\partial P(z,\zeta)}{\partial n}\,ds - \sum_1^k P(z,\alpha_j)\right] \\ &\times\left\{P'(z,\alpha)\prod_1^{m+n-1}\psi(z,\beta_j,\alpha_j)\right. \\ &\qquad\left.\times\exp\left[\sum_1^{m+n-2} P(z,\alpha_j) + \sum_1^m P(z,\beta_j)\right]\right\}^{1/q}; \end{aligned} \tag{6.13}$$

$$\begin{aligned} \omega(z) - \phi^*(z) = C_2 \exp&\left[-\frac{1}{2\pi}\int_\Gamma \ln\rho(\zeta)\frac{\partial P(z,\zeta)}{\partial n}\,ds + \sum_1^k P(z,\alpha_j)\right] \\ &\times\left\{P'(z,\alpha)\prod_1^{m+n-1}\psi(z,\beta_j,\alpha_j)\right. \\ &\qquad\left.\times\exp\left[\sum_1^m P(z,\beta_j) - \sum_1^{m+n-2} P(z,\alpha_j)\right]\right\}^{1/p}, \end{aligned} \tag{6.14}$$

where C, C_1 and C_2 are constants and C_1 is such that

$$\int_\Gamma \left|\frac{f^*(\zeta)}{\rho(\zeta)}\right|^q ds = 1, \qquad q = \infty$$

for the case of Theorem 5.1 (*the class B_ρ^1*), $q = 1$ *for the case of Theorem* 5.6 (*the class $E_{\rho,1}^1$*), *and* $1 < q < +\infty$ *for the case of Theorem* 5.6 (*the class $E_{\rho,q}^1$*), $1/p + 1/q = 1$. *Here $\alpha_1, \dots, \alpha_k$, $k \leqslant m + n - 2$, are zeros of $f^*(z)$ within G.*

PROOF. We use the characteristic property of extremal functions. (We shall argue assuming $q < +\infty$. If $q = +\infty$ (the class B_ρ^1), all computations just simplify.) According to (4.13), we have

$$f^*(\zeta)[\omega(\zeta) - \phi^*(\zeta)]\, d\zeta = \lambda e^{i\delta} \frac{|f^*(\zeta)|^q}{\rho(\zeta)^q}\, ds \tag{6.15}$$

and, hence,

$$e^{-i\delta} R(\zeta)\, d\zeta = e^{-i\delta} f^*(\zeta)[\omega(\zeta) - \phi^*(\zeta)]\, d\zeta \geqslant 0. \tag{6.16}$$

In particular, from (6.16) it follows that $R(z)$ can be analytically continued across Γ (Theorem 5.11). Therefore, we can apply the argument principle to the meromorphic function $R(z)$ with the poles $\beta_1, \dots, \beta_m$:

$$\Delta_\Gamma \arg R(\zeta) = -\Delta_\Gamma \arg d\zeta = n - 2. \tag{6.17}$$

Hence, $R(z)$ has precisely $m + n - 2$ zeros in $\bar{G}$ if zeros on Γ are counted with half-multiplicities. (Multiplicities of such zeros are necessarily even in view of (6.16).) Take the zeros $\alpha_1, \dots, \alpha_{m+n-2}$ of $R(z)$. Taking an arbitrary point $\alpha = \alpha_{m+n-1}$, let us consider the complex-valued Green function $P(z, \alpha)$ with the pole at α. We have

$$\begin{aligned} iP'(\zeta, \alpha) \frac{d\zeta}{ds} &= i \frac{\partial}{\partial s}[P(\zeta, \alpha)] = i - \frac{\partial}{\partial s}[g(\zeta, \alpha) + i\tilde{g}(\zeta, \alpha)] \\ &= -\frac{\partial \tilde{g}(\zeta, \alpha)}{\partial s} = \frac{\partial g(\zeta, \alpha)}{\partial n} > 0. \end{aligned} \tag{6.18}$$

According to (6.18), the function $P'(z, \alpha)$ has $n - 1$ zeros in G. Denote them by $\beta_{m+1}, \dots, \beta_{m+n-1}$. These are the critical points of $g(z, \alpha)$. They are all located within G.

Define a function

$$\mathscr{P}(z) = e^{-i\delta} R(z) / iP'(z, \alpha). \tag{6.19}$$

This is a meromorphic function in $\bar{G}$ satisfying on Γ

$$\mathscr{P}(\zeta) \geqslant 0. \tag{6.20}$$

The points $\beta_1, \dots, \beta_m$, $\beta_{m+1}, \dots, \beta_{m+n-1}$ are the poles of $\mathscr{P}(z)$. The zeros of $\mathscr{P}(z)$ are $\alpha_1, \dots, \alpha_{m+n-2}$, $\alpha = \alpha_{m+n-1}$.

Consider the fraction

$$Q(z) = \frac{\mathscr{P}(z)}{\prod_1^{m+n-1} \psi(z, \beta_j, \alpha_j)}. \tag{6.21}$$

The function $Q(z)$ is analytic in $\bar{G}$ and has constant argument on each of the boundary contours. Therefore, as is easy to see by considering, for example, the mapping $w = Q(z)$, this function is a constant. We have established (6.12).

Let us now justify (6.13). Let $\alpha_1, \dots, \alpha_k$, $k \leqslant m + n - 2$, be those of the zeros of $R(z)$ which are also inner zeros of $f^*(z)$. We wish to show that the harmonic in $\bar{G}$ function

$$u(z) = \ln|f^*(z)| + \sum_1^k g(z, \alpha_j) \tag{6.22}$$

is representable by the Green formula in G.

For this purpose we could use Theorems 5.1, 5.5, and 5.6, but it is even easier to give a direct argument. The classes $E_1(G)$ and $H_1(G)$ coincide in G. The same coincidence holds in the annular domains attached to the boundary contours. Consider the case $1 \leqslant q < +\infty$. (The case $q = \infty$ in a more general situation is covered by Theorem 6.2.) Then, for the conjugate exponent p, we have $1 < p \leqslant \infty$. The function $\omega(z) - \phi^*(z)$ belongs to the class $E_p = H_p \subset D$ in the annuli attached to the boundary contours (see the Appendix). In the same domains, the function $[R(z)]^{-1} \in D$, since $R(z)$ is analytic on Γ. Therefore, in those domains,

$$\frac{1}{f^*(z)} = \frac{\omega(z) - \phi^*(z)}{R(z)} \in D. \tag{6.23}$$

Also,

$$f^*(z) \in E_q(G) = H_q(G) \subset D(G). \tag{6.24}$$

From (6.23) and (6.24) we conclude that $\ln|f^*(z)|$ is representable by the Green formula in each of the annular domains (see the Appendix). The argument is the same as in the proof of the corresponding fact in Theorem 5.8. The function $u(z)$, which differs from $\ln|f^*(z)|$ by a function continuous in those annular domains, is also representable there by the Green integral. Therefore, by Lemma 6.1, such a representation holds for $u(z)$ in G as well.

Taking into account (6.12), (6.7), and (6.15), which hold on Γ, we obtain

$$\begin{aligned} u(z) &= \frac{1}{2\pi}\int_\Gamma u(\zeta)\frac{\partial\phi(\zeta, z)}{\partial n}\,ds \\ &= \frac{1}{2\pi}\int_\Gamma \ln|f^*(\zeta)|\frac{\partial g(\zeta, z)}{\partial n}\,ds \\ &= \frac{1}{2\pi}\int_\Gamma \ln\rho(\zeta)\frac{\partial g(\zeta, z)}{\partial n}\,ds + \frac{1}{2\pi q}\int_\Gamma \ln|P'(\zeta, \alpha)|\frac{\partial g(\zeta, z)}{\partial n}\,ds \\ &\quad + \frac{1}{2\pi q}\sum_1^{m+n-1}\int_\Gamma N(\zeta, \beta_j, \alpha_j)\frac{\partial g(\zeta, z)}{\partial n}\,ds + \ln|C_1|. \end{aligned} \tag{6.25}$$

Compute

$$\frac{1}{2\pi}\int_\Gamma N(\zeta, \beta_j, \alpha_j)\frac{\partial g(\zeta, z)}{\partial n}\,ds. \tag{6.26}$$

Since

$$N(z, \beta_j, \alpha_j) = \ln\left|\frac{z - \alpha_j}{z - \beta_j}\right| + u(z, \beta_j, \alpha_j), \tag{6.27}$$

where $u(z, \beta_j, \alpha_j)$ is harmonic in G and, as is easily seen, is representable by the Green integral, it follows that

$$\frac{1}{2\pi}\int_\Gamma u(\zeta, \beta_j, \alpha_j)\frac{\partial g(\zeta, z)}{\partial n}\,ds = u(z, \beta_j, \alpha_j). \tag{6.28}$$

Furthermore, since $g(z,\alpha_j) = \ln|z-\alpha_j| + u(z,\alpha_j)$, where $u(z,\alpha_j)$ is harmonic in G and representable by Green's formula, we have

$$\frac{1}{2\pi}\int_\Gamma \ln|\zeta-\alpha_j|\frac{\partial g(\zeta,z)}{\partial n}\,ds = \frac{1}{2\pi}\int_\Gamma u(\zeta,\alpha_j)\frac{\partial g(\zeta,z)}{\partial n}\,ds$$
$$= u(z,\alpha_j) = g(z,\alpha_j) + \ln|z-\alpha_j|. \quad (6.29)$$

Similarly,

$$\frac{1}{2\pi}\int_\Gamma \ln|\zeta-\beta_j|\frac{\partial g(\zeta,z)}{\partial n}\,ds = g(z,\beta_j) + \ln|z-\beta_j|. \quad (6.30)$$

Combining (6.28)–(6.30), we obtain

$$\frac{1}{2\pi}\int_\Gamma N(\zeta,\beta_j,\alpha_j)\frac{\partial g(\zeta,z)}{\partial n}\,ds = N(z,\beta_j,\alpha_j) + g(z,\alpha_j) - g(z,\beta_j). \quad (6.31)$$

Substituting this result in (6.25) and using (6.22), we obtain

$$\ln|f^*(z)| = -\sum_1^k g(z,\alpha_j) + \frac{1}{2\pi}\int_\Gamma \ln\rho(\zeta)\frac{\partial g(\zeta,z)}{\partial n}\,ds$$
$$+ \frac{1}{2\pi q}\int_\Gamma \ln|P'(\zeta,\alpha)|\frac{\partial g(\zeta,z)}{\partial n}\,ds$$
$$+ \frac{1}{q}\sum_1^{m+n-1}\left[N(z,\beta_j,\alpha_j) + g(z,\alpha_j) - g(z,\zeta\beta_j)\right] + \ln|C_1|. \quad (6.32)$$

As $\alpha = \alpha_{m+n-1}$ is a pole of $P'(z,\alpha)$ and $\beta_{m+1},\dots,\beta_{m+n-1}$ are zeros of $P'(z,\alpha)$, the function

$$v(z) = \ln|P'(z,\alpha)| + \sum_{m+1}^{m+n-1} g(z,\beta_j) - g(z,\alpha) \quad (6.33)$$

is harmonic in G, continuous in $\overline{G}$, and hence, representable in G by Green's formula. Therefore, we have

$$v(z) = \frac{1}{2\pi}\int_\Gamma \ln|P'(\zeta,\alpha)|\frac{\partial g(\zeta,z)}{\partial n}\,ds. \quad (6.34)$$

Using (6.33) and (6.34) in (6.32), we finally find that

$$\ln|f^*(z)| = -\sum_1^k g(z,\alpha_j) + \frac{1}{2\pi}\int_\Gamma \ln\rho(\zeta)\frac{\partial g(\zeta,z)}{\partial n}\,ds$$
$$+ \frac{1}{q}\ln|P'(z,\alpha)| + \frac{1}{q}\sum_1^{m+n-1} N(z,\beta_j,\alpha_j)$$
$$+ \frac{1}{q}\sum_1^{m+n-2} g(z,\alpha_j) - \frac{1}{q}\sum_1^m g(z,\beta_j) + \ln|C_1|. \quad (6.35)$$

It is clear that (6.35) is equivalent to (6.13). To get (6.14), we simply need to divide (6.12) by (6.13). Note that all factors in (6.13) and (6.14) except the second one are, in general, multivalued. For $q = \infty$, (6.13) becomes (6.1). But we have the estimate $k \leqslant n + m - 2$ for the number k of zeros.

We should mention the following important special case of Theorems 6.2 and 6.3.

THEOREM 6.4. *If under the hypothesis of Theorem* 6.3 *the domain* G *is the unit disk, then*

$$f^*(z)[\omega(z) - \phi^*(z)] = C\frac{\prod_1^m (z-\alpha_j)(1-\bar{\alpha}_j z)}{\prod_1^m (z-\beta_j)(1-\bar{\beta}_j z)}; \tag{6.36}$$

$$f^*(z) = C_1 \exp\left[\frac{1}{2\pi}\int_0^{2\pi} \ln \rho(e^{i\theta}) \frac{e^{i\theta}+z}{e^{i\theta}-z}\, d\theta\right] \prod_1^k \frac{z-\alpha_j}{1-\bar{\alpha}_j z}$$
$$\times \prod_1^{m-1} (1-\bar{\alpha}_j z)^{2/q} \prod_1^m (1-\bar{\beta}_j z)^{-2/q} \quad (k \leqslant m-1); \tag{6.37}$$

$$\omega(z) - \phi^*(z) = C_2 \exp\left[-\frac{1}{2\pi}\int_0^{2\pi} \ln \rho(e^{i\theta}) \frac{e^{i\theta}+z}{e^{i\theta}-z}\, d\theta\right]$$
$$\times \prod_1^{m-1} (1-\bar{\alpha}_j z)^{2/p} \prod_1^m \frac{1-\bar{\beta}_j z}{z-\beta_j} \prod_1^m (1-\bar{\beta}_j z)^{-2/p}, \tag{6.38}$$

$$|\alpha_j| < 1,\ j = 1,\dots,k;\ |\alpha_j| \leqslant 1,\ j = k+1,\dots,m-1.$$

Similarly, if under the hypothesis of Theorem 6.2 *the domain* G *is the unit disk, then* (6.37) *holds for* $f^*(z)$ *with* $q = \infty$, *but the estimate for* k (*the number of zeros of* $f^*(z)$) *is no longer true.*

To get those formulas from (6.12)–(6.14), we have to take into account the fact that for the disk

$$P(z,\alpha) = \ln\frac{z-\alpha}{1-\bar{\alpha}z}, \quad |\alpha| < 1, \qquad \frac{\partial P(z,\zeta)}{\partial n} = \frac{\zeta+z}{\zeta-z},$$

$$P'(z,\alpha) = \frac{1-|\alpha|^2}{(z-\alpha)(1-\bar{\alpha}z)} \neq 0$$

and, therefore, the points $\beta_{m+1},\dots$ do not appear, and

$$\psi(z,\beta_j,\alpha_j) = \frac{(z-\alpha_j)(1-\bar{\alpha}_j z)}{(z-\beta_j)(1-\bar{\beta}_j z)}, \qquad |\alpha_j| \leqslant 1, \quad |\beta_j| \leqslant 1.$$

REMARK 6.5. It should be kept in mind that some of the factors in the representations described by Theorems 6.3 and 6.4 can cancel each other.

Let us point out some special situations which occur quite frequently.

COROLLARY 6.6. *If, under the hypothesis of Theorem* 6.2, *G is the unit disk, then the extremal function in* B^1 *is a finite Blaschke product*:

$$f^*(z) = e^{i\delta}\prod_1^k \frac{z-\alpha_j}{1-\bar{\alpha}_j z}, \qquad |\alpha_j| < 1. \tag{6.39}$$

If, in addition, $\omega(\zeta)$ *is as in Theorem* 6.3, *then* $k \leqslant m-1$.

COROLLARY 6.7. *Under the hypothesis of Theorem* 6.4, *for* $\beta_1 = \beta_2 = \cdots = \beta_m$ *the following polynomials are extremal functions in the class* H_1:

$$\begin{gathered} f^*(z) = C\prod_1^k (z-\alpha_j)(1-\bar{\alpha}_j z) \prod_{k+1}^{m-1} (1-\bar{\alpha}_j z)^2; \\ |\alpha_j| < 1,\ j = 1,\ldots,k; \quad |\alpha_j| \leqslant 1,\ j = k+1,\ldots,m-1. \end{gathered} \tag{6.40}$$

Among those extremal polynomials there always exists a polynomial equal to a square of another polynomial of degree $m-1$ *and not vanishing inside the unit disk.*

(6.40) immediately follows from (6.37). Taking the points α_j, $j = 1,\ldots,k$, and replacing them by arbitrary points α'_j lying on the unit circle, we set

$$\mathscr{F}(z) = C\prod_1^k \left(-\overline{\alpha'_j}\right)\prod_1^k (z-\alpha'_j)^2 \prod_{k+1}^{m-1} (1-\bar{\alpha}_j z)^2. \tag{6.41}$$

$\mathscr{F}(z)/f^*(z) \geqslant 0$ on the unit circle. Therefore, as follows from (4.1), $\mathscr{F}(z)$ is an extremal function as well as $f^*(z)$.

COROLLARY 6.8. *In the extremal problem* (2.13) *for the class* H_1^1 *with an arbitrary continuous* $\omega(\zeta)$, *there exists an extremal function which does not vanish inside the disk.*

For the proof, one has to approximate $\omega(\zeta)$ by the functions $\Sigma\gamma_k/\zeta^k$ and apply the previous corollary.

§7. Mapping properties and zeros of extremal functions

Extremal functions in the class B^1 have the following important mapping property.

THEOREM 7.1. *If under the hypothesis of Theorem* 6.2 *the weight function* $\rho(\zeta) \equiv 1$, *i.e. the extremal problem is posed in the class* B^1 *of functions bounded by one, then the extremal function* $f^*(z)$ *gives a conformal mapping of G onto the k-sheeted unit disk,* $k \geqslant n$, *provided* $f^*(z) \not\equiv e^{i\delta}$. *If the function* $\omega(\zeta)$ *is equal on* Γ *to the boundary values of a meromorphic function in G with m poles, then the following estimate holds for the number k of sheets*:

$$n \leqslant k \leqslant n+m-2. \tag{7.1}$$

PROOF. According to (6.1),

$$f^*(z) = e^{i\delta}\exp\left[-\sum_1^k P(z,\alpha_j)\right], \tag{7.2}$$

where the α_j, $j = 1, \ldots, k$, are zeros of $f^*(z)$. From this it follows that $f^*(z)$ is continuous in $\overline{G}$ and $|f^*(\zeta)| = 1$, $\zeta \in \Gamma$. If $k = 0$, then $f^*(z) \equiv e^{i\delta}$. If $k > 0$, let us map G onto the domain R with an analytic boundary, setting $z = \beta(w)$. It is easy to see that the function $F(w) = f^*(\beta(w))$ is analytic in $\overline{R}$. $F(w)$ maps R onto the whole unit disk K. From the argument principle it follows that each value $t \in K$ is attained by $F(w)$ the same number of times k. Indeed, when a point w traverses the boundary ∂R of R in the positive direction, the point $t = F(w)$ moves along the unit circle in the positive direction. This follows from the fact that we cannot have $F'(w_0) = 0$, $w_0 \in \partial R$. Otherwise, the image of an infinitely small half-neighborhood of w_0 in R would not fit into the unit disk, which contradicts the condition $|f^*(z)| \leqslant 1$. Since to a complete circuit along each boundary contour of R there corresponds a positive integral number of complete circuits of the unit circle, we have that $k \geqslant n$. If $w(z)$ has m poles in G, then the second inequality in (7.1) follows from the estimate of the number of zeros of $f^*(z)$ given in Theorem 6.3.

COROLLARY 7.2. *If under the hypothesis of Theorem* 7.1 *the extremal function* $f^*(z) \equiv \text{const}$, *then the extremal function* $\phi^*(z)$ *is unique.*

Indeed, if $f^*(z) \not\equiv \text{const}$, then $f^*(z)$ has at least n zeros, and it remains to refer to the uniqueness criterion for $\phi^*(z)$ given in Theorem 3.2.

THEOREM 7.3. *Under the hypothesis of Theorem* 6.3, *the function*

$$\mathscr{P}(z) = \frac{e^{-i\delta} f^*(z)[\omega(z) - \phi^*(z)]}{iP'(z, \alpha)} \tag{7.3}$$

maps G onto an s-sheeted plane with slits along the real axis. The number s of sheets satisfies the inequality

$$s \leqslant m + n - 1. \tag{7.4}$$

Here δ is the number which appears in the characteristic property of extremal functions.

PROOF. The assertion of the theorem immediately follows from the fact that $\mathscr{P}(z) \geqslant 0$ on Γ. This was noted in the proof of Theorem 6.3. Estimate (6.4) follows from the fact that $\omega(z)$ has m poles and $P'(z, \alpha)$ has $n - 1$ zeros and, therefore, the number s of poles of $\mathscr{P}(z)$ does not exceed $m + n - 1$. (Some of the poles can cancel with the zeros of $f^*(z)$.)

THEOREM 7.4. *Under the hypothesis of Theorem* 6.3, *the function*

$$\Phi(z) = \int_{z_0}^{z} e^{-i\delta} f^*(t)[\omega(t) - \phi^*(t)]\, dt, \tag{7.5}$$

where z_0 is a fixed point of G, gives a mapping (generally speaking, multivalued) on the plane with slits along the real axis.

This is an easy corollary of the fact that $e^{-i\delta}f^*(\zeta)[\omega(\zeta)-\phi^*(\zeta)]\,d\zeta \geqslant 0$ on Γ.

Information contained in the representations (6.12)–(6.14) of extremal functions can be essentially supplemented by additional information concerning the distribution of their zeros.

THEOREM 7.5. *Under the hypothesis and notation of Theorem* 6.3, *the zeros* $\alpha_1,\dots,\alpha_k$ *of the extremal function* $f^*(z)$ *and the zeros* $\alpha_{k+1},\dots,\alpha_{m+n-2}$ *of the extremal difference* $\omega(z)-\phi^*(z)$ *are connected with the poles* $\beta_1,\dots,\beta_m$ *of* $\omega(z)$ *and critical points* $\beta_{m+1},\dots,\beta_{m+n-1}$ *of the Green function* $g(z,\alpha)$ *by the following relations*:

$$\sum_{r=1}^{m+n-1}\left(\tilde{\omega}_j(\beta_r)-\tilde{\omega}_j(\alpha_r)\right)=\sum_{s=1}^{n-1}\pi_{sj}m_s;$$

$$-\frac{1}{2\pi}\int_\Gamma \ln\rho(\zeta)\frac{\partial\omega_j}{\partial n}\,ds+\sum_{s=1}^{k}\omega_j(\alpha_s)$$
$$+\frac{1}{q}\left[\varepsilon_j-\sum_{s=1}^{m+n-2}\omega_j(\alpha_s)+\sum_{s=1}^{m}\omega_j(\beta_s)\right]\equiv 0\ (\mathrm{mod}\,1) \qquad (7.6)$$

($\alpha=\alpha_{m+n-1}$ *is an arbitrary point of* G). *Here* $j=1,\dots,n-1$, *the* m_s, $s=1,\dots,n-1$ *are certain integers, and* $\varepsilon_j=1$ *if* γ_j *is an inner contour or* $\varepsilon_j=-1$ *if* γ_j *is an outer contour.* $\tilde{\omega}_j$ *is, as usual, a harmonic conjugate of the harmonic measure* ω_j *of* γ_j.

In the opposite direction, the following theorem holds.

THEOREM 7.6. *Let* $\beta_1,\dots,\beta_m$ *and* $\alpha=\alpha_{m+n-1}$ *be given points of* G. *Let* $\beta_{m+1},\dots,\beta_{m+n-1}$ *be critical points of* $g(z,\alpha)$ *and let the points* $\alpha_1,\dots,\alpha_{m+n-2}$ *satisfy* (7.6) *with certain integers* m_s, $s=1,\dots,n-1$, *and* k, $0\leqslant k\leqslant m+n-2$. *There exists a meromorphic function* $\omega(z)$ *in* G *with poles at some of the points* $\beta_1,\dots,\beta_m$ *such that for the extremal problem* (2.20) *the points* $\alpha_1,\dots,\alpha_k$ *are zeros of the extremal function* $f^*(z)$ *and some of the points* $\alpha_{k+1},\dots,\alpha_{m+n-1}$ *are zeros of the extremal difference* $\omega(z)-\phi^*(z)$.

We will omit the proofs.

The following two theorems can be useful for the investigation of extremal functions in the class B^1. (We already know that for those functions $|f^*(\zeta)|=1$, $\zeta\in\Gamma$.)

THEOREM 7.7. *Let the function* $f(z)$ *be analytic in the n-connected domain* $\overline{G}$, *and* $|f(\zeta)|=1$ *on* Γ. *If* $m\geqslant n$ *is the number of zeros of* $f(z)$ *in* G, *then the number of zeros of* $f'(z)$ *is equal to* $m+n-2$.

PROOF. On each of the boundary contours we choose an origin for the arclength and the increase of arclength when traversing in the positive direction. Since $|f(\zeta)|=1$, it follows that $f(\zeta)=e^{i\phi(s)}$. The function $\phi(s)$ is an increasing function of s and, therefore, $d\phi/ds\geqslant 0$. It cannot be equal to zero, since it is impossible that $f'(\zeta) = 0$ (see the proof of Theorem 7.1). We have

$$f'(\zeta) = if(\zeta)\frac{d\phi}{ds}\frac{ds}{d\zeta}. \tag{7.7}$$

Therefore, $\Delta_\Gamma \arg f'(\zeta) = \Delta_\Gamma \arg f(\zeta) - \Delta_\Gamma \arg d\zeta = m + n - 2$.

THEOREM 7.8. *Let G be the unit disk and let $f(z)$ be a Blaschke product (finite an infinite):*

$$f(z) = \prod \frac{\alpha_k - z}{1 - \bar{\alpha}_k z}\frac{\bar{\alpha}_k}{|\alpha_k|}, \qquad |\alpha_k| < 1. \tag{7.8}$$

All zeros of $f'(z)$ in the disk $\{|z| < 1\}$ belong to the smallest convex (in the noneuclidean sense) set F containing all the zeros of $f(z)$.

PROOF. Let $z_0 \notin F$, $|z_0| < 1$. Then there exists a noneuclidean straight line L passing through z_0 such that F lies on one side of L. Consider a fractional-linear transformation of the unit disk onto itself mapping z_0 into 0 and L into the diameter along the real axis. Assume that the image of the set F lies in the upper semidisk, and hence the images $\beta_1, \beta_2, \ldots$ of the points α_1, $\alpha_2, \ldots$ lie in the upper semidisk. The function $f(z)$ goes over into the function $\phi(t)$ of the same form with zeros β_1, $\beta_2, \ldots$. We have

$$\frac{\phi'(t)}{\phi(t)} = \sum \frac{1}{t - 1/\bar{\beta}_k} - \sum \frac{1}{t - \beta_k}.$$

Hence,

$$\frac{\phi'(0)}{\phi(0)} = \sum \bar{\beta}_k \left(\frac{1}{|\beta_k|^2} - 1 \right) \neq 0.$$

Therefore $\phi'(0) \neq 0$. So $f'(z_0) \neq 0$.

§8. The Schwarz lemma for multiply connected domains. Ahlfors', Szegö's, and Green's functions

In this and the following sections we consider applications of the general theory to special extremal problems. The domain G is, as above, n-connected and finite. Its boundary Γ consists of rectifiable Jordan contours γ_i, $i = 1, \ldots, n$. Let $z_0 \in G$. It is natural to call the problem of finding $\sup|f'(z_0)|$ in a certain class of functions analytic in G the *Schwarz lemma problem* in the corresponding class. $B^1(G)$ is, as usual, the class of single-valued functions $f(z)$ in G such that $|f(z)| \leqslant 1$.

THEOREM 8.1. *The extremal function $f^*(z)$ in the problem of*

$$\sup_{f \in B^1(G)} |f'(z_0)| \tag{8.1}$$

is unique up to a factor $e^{i\delta}$ and gives a conformal mapping of G onto the n-sheeted unit disk. In particular, $f^*(z)$ has n zeros, one of which is located at z_0; the others *concide with zeros of the function $\Phi(z)$ analytic in G except at z_0, where it has a simple pole, continuous up to Γ and having a positive differential on Γ:*

$$\Phi(\zeta)\,d\zeta \geqslant 0. \tag{8.2}$$

PROOF. Uniqueness of $f^*(z)$ up to a factor $e^{i\delta}$ follows from Theorem 3.2. The fact that $f^*(z)$ gives a mapping onto the multisheeted disk follows from Theorem 7.1, and the number k of sheets satisfies the estimate

$$n \leqslant k \leqslant n + m - 2, \tag{8.3}$$

where m is the number of poles of the function $\omega(\zeta)$ which defines the extremal problem. According to the Cauchy formula, $\omega(\zeta)$ in problem (8.1) has the form $\omega(\zeta) = 1/2\pi i(\zeta - z_0)^2$ and, therefore, $m = 2$. From (8.3) it follows that $k = n$. If $f^*(z_0) \neq 0$, we should consider the function

$$\phi(z) = \frac{f^*(z) - f^*(z_0)}{1 - \overline{f^*(z_0)}\,f^*(z)},$$

which belongs to $B^1(G)$. Then it is easy to check that $|\phi'(z_0)| > |f^{*\prime}(z_0)|$, which contradicts the fact that f^* is extremal. Now consider the dual problem of (8.1):

$$\inf_{\phi \in E_1(G)} \int_\Gamma \left| \frac{1}{2\pi i(\zeta - z_0)^2} - \phi(\zeta) \right| ds. \tag{8.4}$$

If $\phi^*(z)$ is an extremal function for (8.4), then according to (3.1)

$$f^*(\zeta)\left[\frac{1}{2\pi i(\zeta - z_0)^2} - \phi^*(\zeta)\right] d\zeta \geqslant 0, \tag{8.5}$$

where we have set $e^{i\delta} = 1$ (by choosing $f^*(z)$ suitably). The function

$$\Phi(z) = f^*(z)\left[\frac{1}{2\pi i(z - z_0)^2} - \phi^*(z)\right]$$

has a simple pole at z_0 and is continuous up to Γ. (This follows from §§5 and 6.) It has a positive differential (8.5) on Γ and therefore may have only $n - 1$ zeros in G. All those zeros coincide with the zeros of $f^*(z)$ different from z_0. We shall call the extremal function $f^*(z)$ in problem (8.1), normalized by $f^{*\prime}(z_0) > 0$, the *Ahlfors function* for this problem, and denote it by $f^*(z, z_0)$.

REMARK 8.2. An extremal function in the problem of

$$\sup_{f \in B^1(G)} |f(z_0) - f(z_1)|, \tag{8.6}$$

where z_0 and z_1 are given points in G, also gives an n-sheeted mapping of G on the unit disk.

The extremal problem of the Schwarz lemma for bounded functions solved here is closely related to orthonormal decompositions on the boundary of the domain. Consider the class $E_2(G)$ as a Hilbert space with the inner product

$$(\phi, \psi) = \int_\Gamma \phi(\zeta)\,\overline{\psi(\zeta)}\,ds \tag{8.7}$$

and the norm $\|\phi\| = (\phi, \phi)^{1/2}$.

Let $\{\phi_\nu(z)\}$ be a complete orthonormal system in this space. Show that for each $z_0 \in G$ the series

$$\sum_1^\infty |\phi_\nu(z_0)|^2 < +\infty \tag{8.8}$$

converges. For each $f(z) \in E_2(G)$ the function $f^2(z) \in E_1(G)$ and, therefore,

$$f^2(z_0) = \frac{1}{2\pi i}\int_\Gamma \frac{f^*(\zeta)\,d\zeta}{\zeta - z}.$$

Hence,

$$|f(z_0)|^2 \leqslant \frac{1}{2\pi\rho}\|f\|^2, \tag{8.9}$$

where ρ is the distance of z_0 from Γ. Define the functions

$$K_n(z,t) = \sum_1^n \phi_\nu(z)\overline{\phi_\nu(t)}, \qquad z, t \in G. \tag{8.10}$$

For each n the function $K_n(z,t)$, for t fixed and z varying, belongs to $E_2(G)$. Hence, according to (8.9),

$$\begin{aligned}\left(\sum_1^n |\phi_\nu(z_0)|^2\right)^2 &= |K_n(z_0, z_0)|^2 \leqslant \frac{1}{2\pi\rho}\int_\Gamma |K_n(\zeta, z_0)|^2\,ds \\ &= \frac{1}{2\pi\rho}\int_\Gamma \left|\sum_1^n \phi_\nu(\zeta)\overline{\phi_\nu(z_0)}\right|^2 ds = \frac{1}{2\pi\rho}\sum_1^n |\phi_\nu(z_0)|^2.\end{aligned} \tag{8.11}$$

Dividing by $\Sigma_1^n|\phi_\nu(z_0)|^2$, if this sum is not equal to zero, we obtain

$$\sum_1^n |\phi_\nu(z_0)|^2 \leqslant \frac{1}{2\pi\rho}. \tag{8.12}$$

If $\Sigma_1^n|\phi_\nu(z_0)|^2 = 0$, then (8.12) is evident. (8.8) follows from (8.12). Defining the function

$$K(z,t) = \sum_1^\infty \phi_\nu(z)\overline{\phi_\nu(t)}, \qquad z \in G, \quad t \in G, \tag{8.13}$$

we see that for a fixed t the function $K(z,t)$ is analytic in G. Also, from (8.11) it follows that $K_n(\zeta, t) \to K(\zeta, t)$ in the metric of $E_2(G)$. This implies that $K(z,t) \in E_2(G)$.

Let $f(z)$ be an arbitrary function in $E_2(G)$. Then, on Γ,

$$f(\zeta) = \sum_1^\infty a_n\phi_n(\zeta), \qquad a_n = (f, \phi_n). \tag{8.14}$$

The series in (8.14) converges in the $E_2(G)$-metric. From (8.9) follows convergence of the series

$$f(z) = \sum_1^\infty a_n\phi_n(z) \tag{8.15}$$

to the sum $f(z)$ uniformly on compact subsets of G. We have

$$\begin{aligned}(f(\zeta), K(\zeta, z)) &= \int_\Gamma f(\zeta)\overline{K(\zeta, z)}\, ds \\ &= \int_\Gamma f(\zeta)\cdot \sum_1^\infty \overline{\phi_\nu(\zeta)}\,\phi_\nu(z)\, ds \\ &= \sum_1^\infty a_\nu \phi_\nu(z) = f(z). \qquad (8.16)\end{aligned}$$

In view of (8.16), the function $K(z, t)$ is called the *reproducing Szegö kernel* of the domain G. The Szegö kernel is unique and does not depend on the choice of an orthonormal sequence in the construction. Indeed, let $K_1(z, t)$ be another kernel such that

$$(f, K_1(\zeta, z)) = f(z). \qquad (8.17)$$

Taking the points z and z_0 in G, let us consider the integral

$$\mathcal{I} = \int_\Gamma K(\zeta, z)\overline{K_1(\zeta, z_0)}\, ds. \qquad (8.18)$$

On one hand, according to (8.17), we have

$$\mathcal{I} = K(z_0, z) = \overline{K(z, z_0)}. \qquad (8.19)$$

On the other hand, using (8.16), we find that

$$\bar{\mathcal{I}} = \int_\Gamma K_1(\zeta, z_0)\overline{K(\zeta, z)}\, ds = K_1(z, z_0). \qquad (8.20)$$

Comparing (8.19) and (8.20), we obtain

$$K(z, z_0) \equiv K_1(z, z_0). \qquad (8.21)$$

THEOREM 8.3.

$$\sup_{f\in B^1} |f'(z_0)| = 2\pi K(z_0, z_0). \qquad (8.22)$$

PROOF. Consider the characteristic property of the extremal functions f^* and ϕ^*:

$$f^*(\zeta)\left[\frac{1}{2\pi i(\zeta - z_0)^2} - \phi^*(\zeta)\right] d\zeta = \left|\frac{1}{2\pi i(\zeta - z_0)^2} - \phi^*(\zeta)\right| ds. \qquad (8.23)$$

According to Theorem 8.1, the argument of $f^*(z)$ increases by 2π in going around each boundary contour. In view of (8.23), this implies that the argument of the analytic function

$$1 - 2\pi i(\zeta - z_0)^2\phi^*(\zeta) \qquad (8.24)$$

does not increase in traversing boundary contours. (From the results in §§5 and 6 it is easy to see that $\phi^*(\zeta)$ and, therefore, the function (8.24) are continuous up to

Γ.) Hence, a square root of this function is a single-valued analytic function in G. Set

$$\Phi(z, z_0) = \sqrt{1 - 2\pi i(z - z_0)^2 \phi^*(z)}, \tag{8.25}$$

taking a branch of the root for Φ such that $\Phi(z_0, z_0) = 1$.

Consider the function

$$L(z, z_0) = \frac{\Phi(z, z_0)}{2\pi(z - z_0)}. \tag{8.26}$$

The function $L(z, z_0)$ is analytic in G except at the point z_0, where it has a pole with residue $(2\pi)^{-1}$, and it is continuous up to Γ. Also, set

$$F(z, z_0) = f^*(z) L(z, z_0). \tag{8.27}$$

The function $F(z, z_0)$ is analytic in G, since $f^*(z_0) = 0$, and continuous up to Γ. According to (8.25)–(8.27), (8.23), and the fact that $|f^*(\zeta)| = 1$, (8.23) takes the form

$$F(\zeta, z_0) L(\zeta, z_0)\, d\zeta = i |F(\zeta, z_0)|^2\, ds. \tag{8.28}$$

From this we immediately obtain

$$\overline{F(\zeta, z_0)}\, ds = \frac{1}{i} L(\zeta, z_0)\, d\zeta. \tag{8.29}$$

If we recall the properties of $L(z, z_0)$, then for any $f(z) \in E_2(G)$ it follows from (8.29) that

$$\int_\Gamma f(\zeta) \overline{F(\zeta, z_0)}\, ds = f(z_0). \tag{8.30}$$

Since the function $F(z, z_0)$ has the reproducing property (8.30), it coincides with the Szegö kernel:

$$F(z, z_0) \equiv K(z, z_0). \tag{8.31}$$

Using (8.23)–(8.31), we find that

$$\begin{aligned} \sup_{f \in B^1} |f'(z_0)| &= \int_\Gamma \left| \frac{1}{2\pi i(\zeta - z_0)^2} - \phi^*(\zeta) \right| ds \\ &= 2\pi \int_\Gamma |K(\zeta, z_0)|^2\, ds = 2\pi \int_\Gamma K(\zeta, z_0) \overline{K(\zeta, z_0)}\, ds \\ &= 2\pi K(z_0, z_0). \end{aligned}$$

Thus (8.22) is proved.

REMARK 8.4. The function $L(z, z_0)$, which has a simple pole with residue $(2\pi)^{-1}$ at $z = z_0$ and forms with the Szegö kernel $K(z, z_0)$ the differential (8.29), is also called the *Szegö kernel* (of the second kind). If, in more detail, we denote by $f^*(z, z_0)$ the extremal function $f^*(z)$ for problem (8.1) (the Ahlfors function), then we obtain

$$f^*(z, z_0) = K(z, z_0)/L(z, z_0). \tag{8.32}$$

The Ahlfors function $f^*(z, z_0)$ appears in many estimates and extremal problems for multiply connected domains. (8.32) expresses this function in terms of the Szegö kernel.

Let us consider the complex Green function $P(z, z_0) = g(z, z_0) + i\tilde{g}(z, z_0)$ of G. It is clear that

$$P'(z, z_0) = -1/(z - z_0) + \phi^*(z), \tag{8.33}$$

where $\phi^*(z)$ is analytic in G and continuous in $\overline{G}$. Next, consider all functions $\Phi(z)$ of the form

$$\Phi(z) = -1/(z - z_0) + \phi(z), \tag{8.34}$$

where $\phi(z) \in E_1(G)$. The derivative $P'(z, z_0)$ of the Green function has the following extremal property.

THEOREM 8.5.

$$\inf \int_\Gamma |\Phi(\zeta)|\, ds = \int_\Gamma |P'(\zeta, z_0)|\, ds, \tag{8.35}$$

where the infimum is taken over all Φ *of the form* (8.34).

PROOF. The dual of problem (8.35) is

$$\sup_{f \in B^1} |2\pi f(z_0)|. \tag{8.36}$$

The constants $f^*(z) \equiv e^{i\delta}$ are extremals for this problem. Since $iP'(\zeta, z_0)\, d\zeta \geqslant 0$, we see that

$$iP'(\zeta, z_0)\, d\zeta = |P'(\zeta, z_0)|\, ds. \tag{8.37}$$

According to Theorem 3.2, the functions $\phi^*(z)$ and, therefore, $P'(z, z_0)$ are extremal in problem (8.35). From the argument in §3 it is easy to derive that $\phi^*(z)$ is not a unique extremal in (8.35).

§9. The Schwarz lemma in E_p-classes, $p \geqslant 1$

Let G be an n-connected domain containing ∞ and having a boundary Γ which consists of rectifiable contours $\gamma_1, \ldots, \gamma_n$. Let K be a class of analytic functions in G and let z_0 be an arbitrary point in G. It is convenient to introduce the notation

$$\Omega(\Gamma, K, z_0) = \sup_{f \in K} |f'(z_0)| \tag{9.1}$$

and consider $\Omega(\Gamma, K, z_0)$ as a numerical characteristic of the boundary Γ. If $z_0 = \infty$, then by $f'(\infty)$ we understand

$$f'(\infty) = \lim_{z \to \infty} z[f(z) - f(\infty)] = -\frac{1}{2\pi i}\int_\Gamma f(\zeta)\, d\zeta \tag{9.2}$$

and denote

$$\Omega(\Gamma, K, \infty) = \Omega(\Gamma, K). \tag{9.3}$$

If $K = B^1(G)$, we abbreviate the notation (9.1) and (9.3) by setting

$$\Omega\big(\Gamma, B^1(G), z_0\big) = \Omega(\Gamma, z_0); \qquad \Omega\big(\Gamma, B^1(G)\big) = \Omega(\Gamma). \tag{9.4}$$

If K is a class of analytic functions in G, then by $\tilde{K}$ we denote the subclass of K consisting of functions such that

$$f(\infty) = 0. \tag{9.5}$$

Thus, $f(z) \in \tilde{K}$ has the following expansion in a neighborhood of ∞:

$$f(z) = c_{-1}/z + \cdots. \tag{9.6}$$

By $\tilde{\tilde{K}}$ we denote the subclass of $\tilde{K}$ consisting of functions which have a zero of at least the second order at ∞. So, the inclusion $f(z) \in \tilde{\tilde{K}}$ guarantees the expansion

$$f(z) = c_{-2}/z^2 + \cdots \tag{9.7}$$

in the neighborhood of ∞.

THEOREM 9.1.

$$\Omega\big(\Gamma_1, E_p^1(G)\big) = \Omega\big(\Gamma, \tilde{E}_p^1(G)\big) = (2\pi)^{-1/p}[\Omega(\Gamma)]^{1/q},$$
$$p > 1, 1/p + 1/q = 1; \tag{9.8}$$
$$\Omega\big(\Gamma, E_1^1(G)\big) = \Omega\big(\Gamma, \tilde{E}_1^1(G)\big) \equiv 1/2\pi. \tag{9.9}$$

PROOF. First of all,

$$\Omega(\Gamma) = \Omega\big(\Gamma, B^1(G)\big) = \Omega\big(\Gamma, \tilde{B}^1(G)\big), \tag{9.10}$$

since if (9.5) does not hold for $f(z) \in B^1(G)$, then we define a function

$$\mathcal{F}(z) = \frac{f(z) - f(\infty)}{1 - \overline{f(\infty)}f(z)}$$

for which $|\mathcal{F}'(\infty)| > |f'(\infty)|$ (cf. the proof of Theorem 8.1). Using the duality relations in §2 together with Cauchy's theorem for our domain G containing ∞, we obtain

$$\begin{aligned}\Omega(\Gamma) &= \max_{f \in B^1(G)} \frac{1}{2\pi}\left|\int_\Gamma f(\zeta)\,d\zeta\right| \\ &= \min_{\phi \in \tilde{\tilde{E}}_1(G)} \frac{1}{2\pi}\int_\Gamma |1 + \phi(\zeta)|\,ds;\end{aligned} \tag{9.11}$$

$$\begin{aligned}\Omega\big(\Gamma, \tilde{B}^1(G)\big) &= \max_{f \in \tilde{B}^1(G)} \frac{1}{2\pi}\left|\int_\Gamma f(\zeta)\,d\zeta\right| \\ &= \min_{\phi \in \tilde{E}_1(G)} \frac{1}{2\pi}\int_\Gamma |1 + \phi(\zeta)|\,ds.\end{aligned} \tag{9.12}$$

From (9.10) it follows that

$$\min_{\phi \in \tilde{\tilde{E}}_1(G)} \int_\Gamma |1 + \phi|\,ds = \min_{\phi \in \tilde{E}_1} \int_\Gamma |1 + \phi|\,ds. \tag{9.13}$$

Since the extremal function $f^*(z)$ for the left-hand sides of (9.11) and (9.12) has n zeros in G by Theorem 8.1, the extremal function $\phi^*(z)$ on the right-hand side of (9.12) is unique (see Theorem 3.2). Therefore, according to (9.13), it is also an extremal function for the right-hand side of (9.11). So the common extremal function ϕ^* for the right-hand sides in (9.11) and (9.12) has the expansion (9.7). Set

$$L(z) = 1 + \phi^*(z). \tag{9.14}$$

The characteristic property of the extremal functions f^* and ϕ^* for (9.12) now has the following form:

$$f^*(\zeta)L(\zeta)\,d\zeta = |L(\zeta)|\,ds. \tag{9.15}$$

(Here and in similar cases below we take the extremal function f^* for which $\int_\Gamma f^*(\zeta)\,d\zeta > 0$.) Since (see §8) $\Delta_{\gamma_j} \arg f^*(\zeta) = 2\pi$, $j = 1, \ldots, n$, we have $\Delta_{\gamma_j} \arg L(\zeta) = 0$, and so $\ln L(z)$ is a single-valued analytic function in G. Consider the duality relation

$$\Omega\big(\Gamma, \tilde{E}_p^1(G)\big) = \max_{F \in \tilde{E}_p^1(G)} \frac{1}{2\pi}\left|\int_\Gamma F(\zeta)\,d\zeta\right|$$

$$= \min_{\psi \in \tilde{E}_q(G)} \frac{1}{2\pi}\left\{\int_\Gamma |1 + \psi(\zeta)|^q\,ds\right\}^{1/q}, \qquad p > 1. \tag{9.16}$$

The following equality (Theorem 4.3) is a necessary and sufficient condition for the extremality of $F^*(z)$ and $\phi^*(z)$ for (9.16):

$$F^*(\zeta)[1 + \psi^*(\zeta)]\,d\zeta = \frac{1}{\lambda^{q-1}}|1 + \psi^*(\zeta)|^q\,ds, \tag{9.17}$$

where for brevity we put $\lambda = 2\pi\Omega(\Gamma, \tilde{E}_p^1(G))$, together with the normalizing condition

$$\int_\Gamma |F^*(\zeta)|^p\,ds = 1. \tag{9.18}$$

Note that if (9.17) holds with a certain λ for functions $F^*(z) \in \tilde{E}_p^1(G)$ and $\psi^*(z) \in \tilde{E}_q(G)$ and (9.18) is satisfied, then, necessarily,

$$\lambda = 2\pi\Omega\big(\Gamma, \tilde{E}_p^1(G)\big). \tag{9.19}$$

Set

$$F^*(z) = f^*(z)[L(z)/2\pi\Omega(\Gamma)]^{1/p}, \tag{9.20}$$

where f^* and L are as above, and

$$1 + \psi^*(z) = [L(z)]^{1/q}. \tag{9.21}$$

(We should not forget that $\ln L(z)$ is a single-valued function!) Since $|f^*(\zeta)| = 1$ (Theorem 8.1), we have

$$F^*(\zeta)(1+\psi^*(\zeta))\,d\zeta = \frac{f^*(\zeta)L(\zeta)\,d\zeta}{[2\pi\Omega(\Gamma)]^{1/p}} = \frac{|L(\zeta)|\,ds}{[2\pi\Omega(\Gamma)]^{1/p}} = \frac{|1+\psi^*(\zeta)|^q\,ds}{[2\pi\Omega(\Gamma)]^{1/p}} \tag{9.22}$$

and also

$$\int_\Gamma |F^*(\zeta)|^p\,ds = \int_\Gamma \frac{|L^*(\zeta)|\,ds}{2\pi\Omega(\Gamma)} = 1. \tag{9.23}$$

Comparing (9.22) and (9.23) with (9.17) and (9.18), we conclude that F^* and ψ^* are extremals for (9.16), and that

$$[2\pi\Omega(\Gamma)]^{1/p} = \left[2\pi\Omega\left(\Gamma, \tilde{E}_p^1(G)\right)\right]^{q-1},$$

which coincides with part of (9.8). To prove (9.8) in full, we note that

$$\Omega\left(\Gamma, E_p^1(G)\right) = \max_{F\in E_p^1} \frac{1}{2\pi}\left|\int_\Gamma F(\zeta)\,d\zeta\right| = \min_{\psi\in \tilde{\tilde{E}}_q} \frac{1}{2\pi}\left[\int_\Gamma |1+\psi(\zeta)|^q\,ds\right]^{1/q}. \tag{9.24}$$

But since for $L(z) = 1 + \phi^*(z)$ the function $\phi^*(z)$ has the expansion (9.7), $\psi^*(z)$ in (9.21) has an expansion of the same kind. Therefore the minima in (9.24) and (9.16) are the same. This proves (9.8) in full (see [16]).

Now let us consider the case $p = 1$. The duality relation in this case has the form

$$\Omega\left(\Gamma, \tilde{E}_1^1(G)\right) = \max_{\Phi\in \tilde{E}_1^1} \frac{1}{2\pi}\left|\int_\Gamma \Phi(\zeta)\,d\zeta\right| = \min_{\psi\in \tilde{B}(G)} \frac{1}{2\pi} \max_{\zeta\in\Gamma} |1+\psi(\zeta)|. \tag{9.25}$$

Necessary and sufficient conditions for extremality of Φ^* and ψ^* in (9.25) are

$$\Phi^*(\zeta)[1+\psi^*(\zeta)]\,d\zeta = \lambda|\Phi^*(\zeta)|\,ds, \qquad \lambda > 0; \tag{9.26}$$

$$\int_\Gamma |\Phi^*(\zeta)|\,ds = 1. \tag{9.27}$$

Here, if (9.16) and (9.27) hold, then

$$\lambda = 2\pi\Omega\left(\Gamma, \tilde{E}_1^1(G)\right). \tag{9.28}$$

Since for functions $F(z) \in E_1^1(G)$ the obvious inequality

$$\left|\int_\Gamma F(\zeta)\,d\zeta\right| \leqslant \int_\Gamma |F(\zeta)|\,ds \leqslant 1,$$

holds, it follows that

$$\lambda = 2\pi\Omega\left(\Gamma, E_1^1(G)\right) \leqslant 1. \tag{9.29}$$

On the other hand, from (9.26) it follows that

$$\lambda = |1 + \psi^*(\zeta)|. \tag{9.30}$$

For $z = \infty$ we have $1 + \psi^*(\infty) = 1$. So, according to the maximum modulus principle for analytic functions, $\lambda \geqslant 1$. Therefore, $\lambda = 1$ and $\psi^*(z) \equiv 0$. Now it is clear that (9.9) holds. The proof is complete.

We leave it to the reader to prove by an argument similar to the one we used above in Theorem 9.1, the following theorem.

THEOREM 9.2. *For an arbitrary point $z_0 \in G$,*

$$\max_{F \in \tilde{E}_2^1(G)} |F(z_0)| = \left[\frac{\Omega(\Gamma, z_0)}{2\pi}\right]^{1/2}. \tag{9.31}$$

§10. Some extremal problems for bounded analytic functions in the disk

In this section we consider the class B^1 in the unit disk. We illustrate the general theory of §§1–7 by investigating the problems which historically were the first examples of nontrivial extremal problems for bounded analytic functions. A thorough analysis of the solutions of those problems has led to the idea of duality for extrema in the theory of analytic functions. Let c_k denote the Taylor coefficient at z^k of a function $f(z)$ analytic in the unit disk.

THEOREM 10.1. *Assume that the polynomial*

$$P(z) = \lambda_0 + \lambda_1 z + \cdots + \lambda_n z^n \tag{10.1}$$

does not have zeros in the disk $|z| \leqslant 1$. Let

$$[P(z)]^2 = \mu_0 + \mu_1 z + \cdots + \mu_n z^n + \cdots. \tag{10.2}$$

Then

$$\sup_{f \in B^1} |\mu_n c_0 + \mu_{n-1} c_1 + \cdots + \mu_0 c_n| = \sum_{j=0}^{n} |\lambda_j|^2. \tag{10.3}$$

An extremal function for (10.3) *is unique up to a factor $e^{i\delta}$ and has the form*

$$f^*(z) = e^{i\delta} \frac{\bar{\lambda}_n + \bar{\lambda}_{n-1} z + \cdots + \bar{\lambda}_0 z^n}{\lambda_0 + \lambda_1 z + \cdots + \lambda_n z^n}. \tag{10.4}$$

PROOF. For $f \in B^1$ we have

$$\begin{aligned} |\mu_n c_0 + \cdots + \mu_0 c_n| &= \left|\frac{1}{2\pi i}\int_\Gamma \frac{f(\zeta)\,d\zeta}{\zeta^{n+1}}(\mu_0 + \cdots + \mu_n \zeta^n)\right| \\ &= \left|\frac{1}{2\pi i}\int_\Gamma \frac{f(\zeta)\,d\zeta}{\zeta^{n+1}}(\lambda_0 + \cdots + \lambda_n \zeta^n)^2\right| \\ &\leqslant \frac{1}{2\pi}\int_\Gamma |\lambda_0 + \cdots + \lambda_n \zeta^n|^2\,ds = \sum_0^n |\lambda_j|^2. \end{aligned} \tag{10.5}$$

On the other hand, by a direct calculation we verify that for $f^*(z)$ of the form (10.4), equalities occur everywhere in the chain of inequalities (10.5). Uniqueness of the extremal function $f^*(z)$ follows from the general theory in §3 or can be verified directly by analyzing (10.5).

REMARK 10.2. According to Corollary 6.6, the extremal $f^*(z)$ must be a finite Blaschke product. It is easy to see that (10.4) is a Blaschke product. Each finite Blaschke product can be written in the form (10.4) with appropriate λ_j.

LEMMA 10.3. *Let the numbers* $\{\lambda_j\}$, $j = 0, \ldots, n$ *satisfy the inequalities*

$$\lambda_0 > \lambda_1 > \cdots > \lambda_n > 0. \tag{10.6}$$

Then the polynomial

$$P(z) = \lambda_n z^n + \lambda_{n-1} z^{n-1} + \cdots + \lambda \tag{10.7}$$

has no zeros in the disk $|z| \leqslant 1$.

PROOF. For $|z| \leqslant 1$ and $z \neq 1$, we have

$$\begin{aligned} &|(1-z)P(z)| \\ &\quad = |\lambda_0 - (\lambda_0 - \lambda_1)z - (\lambda_1 - \lambda_2)z^2 - \cdots - (\lambda_{n-1} - \lambda_n)z^n - \lambda_n z^{n+1}| \\ &\quad \geqslant \lambda_0 - |(\lambda_0 - \lambda_1)z + \cdots + \lambda_n z^n|. \end{aligned} \tag{10.8}$$

If $z \geqslant 0$, it is clear that $P(z) \neq 0$. If z is not a positive number, then all the powers $z, z^2, \ldots, z^{n+1}$ cannot have the same argument and, therefore, from (10.8) we obtain a strict inequality:

$$|(1-z)P(z)| > \lambda_0 - (\lambda_0 - \lambda_1) - \cdots - \lambda_n = 0.$$

This proves the lemma.

COROLLARY 10.4. *If the coefficients of the polynomial* (10.7) *satisfy* (10.6) *and the numbers* $\mu_0, \ldots, \mu_n$ *are defined by* (10.2), *then*

$$\sup_{f \in B^1} |\mu_n c_0 + \cdots + \mu_0 c_n| = \sum_{j=0}^{n} \lambda_j^2 \tag{10.9}$$

and the unique (up to a factor) extremal function has the form (10.4), *where now, obviously,* $\overline{\lambda}_j = \lambda_j$.

COROLLARY 10.5.

$$\begin{aligned} &\sup_{f \in B^1} |c_0 + c_1 + \cdots + c_n| \\ &\quad = 1 + \left(\frac{1}{2}\right)^2 + \left(\frac{1 \cdot 3}{2 \cdot 4}\right)^2 + \cdots + \left(\frac{1 \cdot 3 \cdot \cdots \cdot (2n-1)}{2 \cdot 4 \cdot \cdots \cdot 2n}\right)^2 \end{aligned} \tag{10.10}$$

and the unique (up to a factor $e^{i\delta}$) extremal function $f^*(z)$ has the form

$$f^*(z) = e^{i\delta}\frac{\dfrac{1\cdot 3\cdot\dots\cdot(2n-1)}{2\cdot 4\cdot\dots\cdot 2n} + \dots + \dfrac{1\cdot 3}{2\cdot 4}z^{n-2} + \dfrac{1}{2}z^{n-1} + z^n}{1 + \dfrac{1}{2}z + \dots + \dfrac{1\cdot 3\cdot\dots\cdot(2n-1)}{2\cdot 4\cdot\dots\cdot 2n}z^n}. \tag{10.11}$$

PROOF. Consider the series

$$(1-z)^{-1/2} = 1 + \frac{1}{2}z + \frac{1\cdot 3}{2\cdot 4}z^2 + \dots + \frac{1\cdot 3\cdot\dots\cdot(2n-1)}{2\cdot 4\cdot\dots\cdot 2n}z^n + \dots. \tag{10.12}$$

Comparing (10.12) with the geometric series $(1-z)^{-1}$, we find that

$$P_n^2(z) = \left(1 + \frac{1}{2}z + \dots + \frac{1\cdot 3\cdot\dots\cdot(2n-1)}{2\cdot 4\cdot\dots\cdot 2n}z^n\right)^2 = 1 + z + \dots + z^n + \dots. \tag{10.13}$$

It remains to refer to the previous corollary.

Corollary 10.5 (a theorem of Landau [37]) was probably the first nontrivial extremal problem for bounded analytic functions in the disk. This problem is quite natural, since it concerns

$$\sup_{f\in B^1}\max_{|z|\leqslant 1}|S_n(f,z)|,$$

where $S_n(f,z)$ is the nth partial sum of the Taylor expansion of f computed at z.

§11. The role of Ahlfor's function in other extremal problems

In this section we will show the role of the Ahlfors function in extremal problems of Schwarz lemma type which are posed in other classes of analytic functions, as well as in other extremal problems.

Let G be an n-connected bounded domain with nondegenerating boundary continua. For $a \in G$, let us denote by $B_a(G)$ the subclass of $B^1(G)$ consisting of functions for which

$$f(a) = 0. \tag{11.1}$$

Let $F(z,a)$ be the Ahlfors function. It belongs to $B_a(G)$. We shall say that a meromorphic function $f(z)$ in G is *subordinate* in G, with respect to a point $a \in G$, to a function $g(\tau)$ meromorphic in the disk $|\tau| < 1$ if

$$f(z) = g(\omega(z)), \tag{11.2}$$

where $\omega(z) \in B_a(G)$; we write $f(z) \prec_a g(\tau)$. The following theorem is evident (see [2]).

THEOREM 11.1 *If $f(z) \prec_{z_0} g(\tau)$ and $f(z)$ is regular at z_0, then*

$$|f'(z_0)| \leqslant |g'(0)|\,|F'(z_0,z_0)|, \tag{11.3}$$

and for $g'(0) \neq 0$ *equality is attained only if*

$$f(z) = g(\varepsilon F(z, z_0)), \qquad |\varepsilon| = 1. \tag{11.4}$$

COROLLARY 11.2. *Fix* $f(z) \in B^1(G)$ *and for all* $z_0 \in G$,

$$|f'(z_0)| \leqslant \left(1 - |f(z_0)|^2\right) F'(z_0, z_0). \tag{11.5}$$

Equality in (11.5) *is attained if and only if*

$$f(z) = \varepsilon \frac{F(z, z_0) + \alpha}{1 + \bar{\alpha} F(z, z_0)}, \qquad |\alpha| < 1, \quad |\varepsilon| = 1. \tag{11.6}$$

Indeed, if $f(z) \in B^1(G)$, then the function

$$\omega(z) = \frac{f(z) - \alpha}{1 - \bar{\alpha} f(z)}, \qquad \alpha = f(z_0),$$

belongs to B_{z_0}. Since $f(z) = (\omega(z) + \alpha)/(1 + \bar{\alpha}\omega(z))$, it follows that $f(z) \prec_{z_0} g(\tau)$, where $g(\tau) = (\tau + \alpha)/(1 + \bar{\alpha}\tau)$. Applying (11.3) we obtain (11.5). Equality is attained in (11.5) only if $\omega(z) = \varepsilon F(z, z_0)$, i.e., when $f(z)$ is as in (11.6). Conversely, for $f(z)$ of the form (11.6) with arbitrary α, $|\alpha| < 1$, and ε, $|\varepsilon| = 1$, equality holds in (11.5).

Inequality (11.5) strengthens in a certain way the inequality $|f'(z_0)| \leqslant F'(z_0, z_0)$ for an arbitrary $f \in B^1(G)$.

COROLLARY 11.3. *If a function* $f(z)$ *is analytic in* G *and*

$$\operatorname{Re} f(z) > 0, \tag{11.7}$$

then

$$|f'(z_0)| \leqslant 2 \operatorname{Re} f(z_0) F'(z_0, z_0) \tag{11.8}$$

and equality holds in (11.8) *for the functions*

$$f(z) = \beta \frac{1 + \varepsilon F(z, z_0)}{1 - \varepsilon F(z, z_0)} + i\gamma, \qquad |\varepsilon| = 1, \quad \beta > 0, \tag{11.9}$$

where γ *is real.*

COROLLARY 11.4. *If a function* $f(z)$ *is regular in* G *and*

$$|\operatorname{Re} f(z)| \leqslant 1, \tag{11.10}$$

then

$$|f'(z_0)| \leqslant \frac{4}{\pi} F'(z_0, z_0), \tag{11.11}$$

and equality holds in (11.11) *only for the functions*

$$f(z) = \frac{2}{\pi i} \ln \beta \frac{1 + \varepsilon F(z, z_0)}{1 - \varepsilon F(z, z_0)}. \tag{11.12}$$

It is convenient to reduce the proofs of Corollaries 11.3 and 11.4 to Corollary 11.2 rather than directly to Theorem 11.1. Let us prove Corollary 11.3. The function $\phi(z) = (f(z) - 1)/(f(z) + 1)$ belongs to $B^1(G)$, and

$$\frac{|\phi'(z)|}{1 - |\phi(z)|^2} = \frac{|f'(z)|}{2\operatorname{Re} f(z)}. \tag{11.13}$$

Using (11.5), we obtain (11.8). Equality holds in (11.5) only for $\phi(z)$ of the form (11.6), and this leads to (11.9) for $f(z)$. Similarly, one can prove Corollary 11.4. Now the function $\psi(z)$ defined by

$$\frac{\psi(z) + 1}{\psi(z) - 1} = e^{i\pi f(z)/2} \tag{11.14}$$

belongs to the class $B^1(G)$.

COROLLARY 11.5. *If a function $f(z)$ belongs to $B^1(G)$ and its logarithm is a single-valued analytic function in G, then*

$$|f'(z_0)| \leqslant \frac{2}{e} F'(z_0, z_0) \tag{11.15}$$

and equality only holds for functions of the form

$$f(z) = \eta \exp\left[-\frac{1 + \varepsilon F(z, z_0)}{1 - \varepsilon F(z, z_0)}\right], \qquad |\varepsilon| = |\eta| = 1. \tag{11.16}$$

PROOF. The function $\chi(z) = -\ln f(z)$ has a positive real part in G. Hence,

$$|f'(z_0)| \leqslant |f(z_0)|2\ln\frac{1}{|f(z_0)|}F'(z_0, z_0). \tag{11.17}$$

The function $\Phi(x) = -x\ln x$ has a maximum at $x = 1/e$, and this yields (11.15). Equality in (11.15) can occur if and only if $\chi(z)$ has the form (11.9) with $\beta = 1$. This leads us to the formula (11.16) for $f(z)$. Consider the class $\mathscr{B}_a(G)$ consisting of single-valued analytic functions in G for which the area of the values w omitted by the function $w = 1/(f(z) - f(a))$ is greater or equal to π, i.e. $f(z) \in \mathscr{B}_a(G)$, if the area of A is at least π, where

$$A = \{w: \nexists z \in G, 1/(f(z) - f(a)) = w\}. \tag{11.18}$$

It is clear that $B_a(G) \subset \mathscr{B}_a(G)$.

THEOREM 11.6.

$$\sup_{f \in \mathscr{B}_a(G)} |f'(a)| = F'(a, a). \tag{11.19}$$

PROOF. Let $f(z) \in \mathscr{B}_a(G)$. Considering problem (11.19), we can assume that $f(a) = 0$. Consider an analytic function $F(z)$ in G,

$$F(z) = \frac{1}{Z^2(A)} \iint_A \frac{d\sigma_w}{1/f(z) - w}, \tag{11.20}$$

where $d\sigma_w$ is an area element and $Z^2(A)$ is the area of A. As one can easily compute, we have

$$F'(a) = f'(a). \tag{11.21}$$

The theorem will be proved if we show that $|F(z)| \leqslant 1$. For this purpose we will show that for all t

$$\left| \iint_A \frac{d\sigma_w}{t - w} \right| \leqslant Z^2(A). \tag{11.22}$$

By simple substitutions we reduce the estimate of the integral in (11.22) to the estimate of the following integral:

$$\mathcal{I} = \iint_A \frac{d\sigma}{\zeta}. \tag{11.23}$$

(We can assume that the value of the integral is positive.) Denote by A^+ the part of A which belongs to the right half-plane of the variable ζ. Introducing the polar coordinates (r, θ), we obtain

$$\mathcal{I} \leqslant \iint_{A^+} \cos\theta \, dr \, d\theta. \tag{11.24}$$

Let $l(r, \theta)$ denote the linear measure of the set $\{\zeta \in A^+ \colon \arg\zeta = \theta \text{ and } |\zeta| \leqslant r\}$. Put $l(\theta) = l(\infty, \theta)$. Then, applying the Schwarz-Bunyakovskiĭ inequality, we obtain

$$\iint_{A^+} \cos\theta \, dr \, d\theta = \int_{-\pi/2}^{\pi/2} l(\theta) \cos\theta \, d\theta \leqslant \left(\frac{\pi}{2} \int_{-\pi/2}^{\pi/2} l^2(\theta) \, d\theta \right)^{1/2}. \tag{11.25}$$

On the other hand, it is clear that $l(r, \theta) \leqslant r$ and $dl(r, \theta) \leqslant dr$. Therefore,

$$\int_{A_\theta^+} r \, dr \geqslant \int_{A_\theta^+} l(r, \theta) \, dl(r, \theta) = \frac{l^2(\theta)}{2}. \tag{11.26}$$

$(A_\theta^+ = \{w \colon w \in A^+, \arg w = \theta\}$.) Hence,

$$Z^2(A) \geqslant Z^2(A^+) = \int_{-\pi/2}^{\pi/2} d\theta \int_{A_\theta^+} r \, dr \geqslant \frac{1}{2} \int_{-\pi/2}^{\pi/2} l^2(\theta) \, d\theta. \tag{11.27}$$

Combining (11.24), (11.25), and (11.27), we arrive at the desired estimate

$$\mathcal{I} \leqslant \left[\pi Z^2(A)\right]^{1/2} \leqslant Z^2(A). \tag{11.28}$$

Let $C_a(G)$ be the class of single-valued analytic functions in G satisfying the conditions

$$f(a) = 0, \qquad f(z_1) \cdot f(z_2) \neq 1, \quad \forall z_1, z_2 \in G. \tag{11.29}$$

In particular, when G is the unit disk and $a = 0$, the class C_a coincides with the class introduced by Bieberbach and Eilenberg—see [4]. The following inclusions hold:

$$B_a(G) \subset C_a(G) \subset \mathcal{B}_a(G). \tag{11.30}$$

The first one is clear. Let us prove the second. Let A be the image of G under the mapping $w = 1/f(z)$, $f(z) \in C_a(G)$. It is obvious that $A \ni \infty$. Let A_θ be the intersection of A with the ray issuing from the origin and having argument θ. Let B_θ be a complement of A_θ on this ray. The area $Z^2(B)$ of the complement of A is given by

$$Z^2(B) = \int_{-\pi}^{\pi} d\theta \int_{B_0} r\,dr. \tag{11.31}$$

(11.29) shows that if $w_1, w_2 \in A$, then $w_1 w_2 \neq 1$. Therefore, the set $B_{-\theta}$ contains the image $\tilde{A}_\theta$ of A_θ under the mapping $r' = 1/r$. (We assume that all the A_θ and B_θ belong to $[0, +\infty)$.) Therefore,

$$Z^2(B) \geqslant \int_0^{\pi} \left(\int_{B_\theta} r\,dr + \int_{\tilde{A}_\theta} r\,dr \right) = \int_0^{\pi} \left(\int_{B_\theta} r\,dr + \int_{A_\theta} \frac{dr}{r^3} \right). \tag{11.32}$$

Denote by B_θ^1 and A_θ^1 the intersections of B_θ and A_θ with the segment $[0, 1]$, and by B_θ^2 and A_θ^2 the intersections of those sets with $(1, +\infty)$. Then

$$\int_{B_\theta^1} r\,dr + \int_{A_\theta^1} \frac{dr}{r^3} \geqslant \int_0^1 r\,dr = \frac{1}{2}; \tag{11.33}$$

$$\int_{B_\theta^2} r\,dr + \int_{A_\theta^2} \frac{dr}{r^3} \geqslant \int_1^{\infty} \frac{dr}{r^3} = \frac{1}{2}. \tag{11.34}$$

Combining (11.32)–(11.34), we find that $Z^2(B) \geqslant \pi$, and the inclusion (11.30) is verified.

COROLLARY 11.7.

$$\sup_{f \in C_a(G)} |f'(a)| = F'(a, a). \tag{11.35}$$

Certainly, this is not a complete list of problems where the Ahlfors function is involved in the solution.

Historical notes

§1. M. G. Kreĭn's paper [33] should be considered as the beginning of the systematic use of duality relations following from the Hahn-Banach theorem in the investigation of problems on best approximation. In S. M. Nikol'skiĭ's paper [44] the duality relations were established in a slightly different form (Theorems 1.1 and 1.2). Theorem 1.2 was proved in [44] for a finite-dimensional E, and Theorem 1.1 for E with finite codimension. At the present time there is a tremendous amount of literature where the duality method is substantially generalized, constituting an important part of mathematical (above all, convex) programming and is applied to a great variety of extremal problems. In particular, a wide circle of problems of approximation theory essentially using duality arguments is presented in the monographs by Voronovskaya [11], Gol'shteĭn [23], Korniĭchuk [31], Kreĭn and Nudel'man [36], V. M. Tikhomirov [59], and Laurent [38].

§**2.** The extremal problem of $\sup|\int_\Gamma f(\zeta)\omega(\zeta)\,d\zeta|$ studied here is a natural generalization of problems of the type

$$\sup\left|\sum_{k=1}^{r}\cdot\sum_{j=0}^{m_k}\alpha_{kj}f^{(j)}(z_k)\right| \qquad (*)$$

(the z_k are given points, and the α_{kj} are given coefficients), which are the subject of an extremely vast literature. The first extremal problem of this type was Landau's problem in [37] on $\sup|c_0+\cdots+c_n|$ in the class B^1 in the unit disk (see Corollary 10.5). The generalization of Landau's problem is the Szász problem in [55] (Theorem 10.1). Since then this problem in the class B^1 in the unit disk has been considered various special α_{kj} by many authors. F. Riesz, in [50], considered the problem of $\inf\int_0^{2\pi}|\phi(\zeta,\alpha_m,\dots,\alpha_0)|\,d\theta$ in the class H_1, where the infimum is taken over functions with a given nth Taylor polynomial $\alpha_m+\alpha_{m-1}z+\cdots+\alpha_0 z^m$. In [56], Szász established the duality relation

$$\sup_{f\in B^1}\left|\sum_0^m\alpha_k c_k\right| = \inf_{H_1}\frac{1}{2\pi}\int_0^{2\pi}|\phi(\zeta,\alpha_m,\dots,\alpha_0)|\,d\theta.$$

His proof of the relation was based on the special algebraic structure of the extremal function ϕ^* found by Riesz. Riesz himself used a duality relation of a somewhat different kind in his investigation. Szász' result was independently repeated by Geronimus in [17]–[20] and, later, by Goluzin in [22]. Geronimus and Goluzin also established a relation dual to the above (interchanging the roles of B^1 and H^1 and replacing the integral metric on the right-hand side by the uniform metric). The most detailed survey of the results related to the problem $(*)$ in the disk for all classes H_p^1, $p\geqslant 1$, is given in [41] by Macintyre and Rogosinski; their paper also contains a long list of references. The study of problem $(*)$ for multiply-connected domains began much later. The paper of Grunsky [24], where he was studying the problem of $\sup|f(b)|$, $f\in B^1$, $f(a)=0$ (Remark 8.2), was the first in that area. Ahlfors in [5] studied a related problem concerning $\sup|f'(a)|$, $f\in B^1$, $f(a)=0$, and obtained significantly more complete results. Garabedian in [15] established the duality relations for the Ahlfors problem. Robinson [51] and later Heins [72], almost simultaneously with Grunsky, considered extremal problems in the circular annulus.

In all these papers, the basis for establishing the duality relations lay in special properties of extremal functions. The general nature of the duality relations had not yet been revealed, and they had not been a tool for investigation of extremal functions, but, on the contrary, were derived from the properties of those functions. The study of the duality of extremal problems in the theory of analytic functions, based on the Hahn-Banach theorem and, first of all, Theorem 1.1, was started by Khavinson in [63], [64]. Earlier, as we have already mentioned, Kreĭn [33] and Nikol'skiĭ [44] applied a similar method to approximation problems. Following [63] and [64] there have appeared a large number of articles by various authors studying the duality of extremal problems in the theory of analytic functions. We mention papers by Rogosinski and Shapiro [52], Bonsall [8], Rudin [54], Lax [40], Penez [46], Royden [53], Read [48], [49], Khavinson [65], [66], [68], Tumarkin and Khavinson [60], Kochetkov [32], Hejhal [73], [74], Gamelin [14], and Watanabe [9]. The further history of the problem and a very extensive bibliography can be found in the surveys [19], [41], [43], [61] and [69].

In §2 and further in §§3–4 we follow mainly [66].

§§3 and 4. Doob was the first, in [26], to study L_1-approximation of an arbitrary function on the unit circle by the boundary values of analytic functions. His method was not based on duality. The example with a nonexisting extremal $f^*(z)$ in Theorem 4.2 is given in [52] and [66]. The example of where an extremal function $\phi^*(z)$ is not unique is taken from [29]. Further interesting results concerning properties of the extremal functions $f^*(z)$ and $\phi^*(z)$ in Theorem 4.2 assuming only the continuity of $\omega(\zeta)$ are contained in de Leeuw and Rudin's paper [39], in Adamyan, Arov and Kreĭn [34], [35], and in Carleson and Jacobs [28].

§5. We follow [60], where the original method of [66] has been significantly strengthened. However, we present the method under more stringent restrictions than in [60] to avoid technical difficulties.

§6. The main Theorem 6.3 is due to the author [71]. A partial case of Theorem 6.3 related to the Schwarz lemma for H_∞ and E_2 is contained in [10] and, in a different form, in an earlier paper [15]. Corresponding results for the disk for $\rho(\theta) \equiv 1$ (Theorem 6.4) have been known for a long time, and have been obtained by many authors. The most detailed presentation for the case of the disk can be found in [41], [52], and [19]. In the case of a circular annulus, a representation of the extremal functions is given in Voskanyan's paper [12]. In [58], formulas (6.36)–(6.38) are extended to the Orlicz classes generalizing the classes H_p, $p \geqslant 1$. Corollary 6.8 was noted by Goluzin in [21]. Theorem 7.1 is due to Khavinson—see [65].

§7. Partial cases of Theorem 7.1 have been earlier considered by Ahlfors [5], Grunsky [24], and Garabedian [15] (see §8). Theorems 7.3 and 7.4 were mentioned in those papers for more specific problems. For the special case of the Schwarz lemma (see §8), Theorems 7.5 and 7.6 are contained in [10]. The general case is considered in [71]. Theorem 7.7 for the unit disk is due to Macdonald. Theorem 7.8 is due to Walsh; his monograph [62] deals with questions associated with these two theorems.

§8. Theorem 8.1 is due to Ahlfors [5]. Remark 8.2 is connected with Grunsky's extremal problem [24]. Theorem 8.3 was proved by Garabedian [15]. To our knowledge, Theorem 8.5 has not been noted before. For the circular annulus, the constructive solution (in terms of elliptic functions) was obtained for a number of extremal problems of the type considered in this section. We mention the papers by Robinson [51], Alenitsyn [2], and Voskanyan [13]. Also, we mention efficient solutions of some interesting extremal problems obtained by Osipenko in [45].

§9. Theorems 9.1 and 9.2 were established by Garabedian in [16] and somewhat generalized in [61].

§10. Theorem 10.1 is due to Szász [55]. Lemma 10.3 is a known theorem of Eneström and Kakeya (see, for instance, [47], problem 22 in Part III). The Landau estimate (10.10) has been generalized by many authors. Stechkin [57] considered the problem of estimating the remainder of the Taylor series for functions of the class B^r for which $|f^{(r)}(z)| \leqslant 1$ (r is a natural number). Babenko [7] simplified Stechkin's argument, and Markushevich and Khavinson [70] considered an estimate of Taylor sums centered at the point $\alpha \neq 0$ for the class B^1. In [42], Mordasova extended those results to the classes B^r. In particular, (10.11)-type estimates can be used to study problems concerning over-convergence of power series.

§11. The concept of subordination used in this section was introduced by Alenitsyn in [3]. It is a natural generalization of the well-known subordination in the unit disk (see, for instance, [21]). Results 11.1–11.5 are mentioned in [1] and [3]. Theorem 11.6 was proved by Ahlfors and Beurling [6].

References

1. Yu. E. Alenitsyn, *On bounded functions in multiply connected domains*, Dokl. Akad. Nauk SSSR **73** (1950), 245–248. (Russian)

2. ______, *On some estimates for functions regular in a finitely-connected domain*, Mat. Sb. **49 (91)** (1959), 181–190. (Russian)

3. ______, *An extension of the principle of subordination to multiply connected domains*, Trudy Mat. Inst. Steklov. **60** (1961), 5–21; English transl. in Amer. Math. Soc. Transl. (2) **43** (1964).

4. Yu. E. Alenitsyn, G. V. Kuz'mina and N. A. Lebedev, *Methods and results of geometric function theory*, Supplement to G. M. Goluzin, *Geometric theory of functions of a complex variable*, 2nd ed., "Nauka", Moscow, 1966, pp. 532–626; English transl., Amer. Math. Soc., Providence, R. I., 1969, pp. 563–672.

5. Lars V. Ahlfors, *Bounded analytic functions*, Duke Math. J. **14** (1947), 1–11.

6. Lars Ahlfors and Arne Beurling, *Conformal invariants and function-theoretic null-sets*, Acta Math. **83** (1950), 101–129.

7. K. I. Babenko, *On the best approximation to a class of analytic functions*, Izv. Akad. Nauk SSSR Ser. Mat. **22** (1958), 631–640. (Russian)

8. F. F. Bonsall, *Dual extremum problems in the theory of functions*, J. London Math. Soc. **31** (1956), 105–110.

9. Takeshi Watanabe, *Reciprocal extremal problems in conformal mapping*. I, Bull. Fac. Liberal Arts Ibaraki Univ. (Nat. Sci.) **1959**, no. 10, 17–24. R. Zh. Mat. **1962** #5Б155.

10. Harold Widom, *Extremal problems associated with a system of curves in the complex plane*, Advances in Math. **3** (1969), 127–232.

11. E. V. Voronovskaya, *The functional method and its applications*, Leningrad. Èlektrotekhn. Inst. Svyazi, Leningrad, 1963; English transl., Amer. Math. Soc., Providence, R. I., 1970.

12. V. V. Voskanyan, *A certain extremal problem in the theory of Banach spaces of functions that are analytic in an annulus*, Izv. Akad. Nauk Armyan. SSR Ser. Mat. **6** (1971), 412–418. (Russian)

13. ______, *The values largest in modulus for certain classes of functions analytic in an annulus*, Akad. Nauk Armyan. SSR Dokl. **55** (1972), 203–206. (Russian)

14. T. W. Gamelin, *Extremal problems in arbitrary domains*, Michigan Math. J. **20** (1973), 3–11.

15. P. R. Garabedian, *Schwarz's lemma and the Szegö kernel function*, Trans. Amer. Math. Soc. **67** (1949), 1–35.

16. ______, *The classes L_p and conformal mapping*, Trans. Amer. Math. Soc. **69** (1950), 392–415.

17. Ya. L. Geronimus, *Sur le problème des coefficients pour les fonctions bornées*, C. R. (Dokl.) Acad. Sci. URSS **14** (1937), 97–98.

18. ______, *Sur quelques problèmes extrémaux*, Izv. Akad. Nauk SSSR Ser. Mat. **1** (1937), 185–202. (Russian; French summary)

19. ______, *On certain extremal properties of analytic functions*, Izv. Akad. Nauk SSSR Ser. Mat. **12** (1948), 325–336. (Russian)

20. ______, *Sur un problème de F. Riesz et le problème généralisé de Tchebycheff-Korkine-Zolotareff*, Izv. Akad. Nauk SSSR Ser. Mat. **3** (1939), 279–288. (Russian; French summary)

21. G. M. Goluzin, *Estimates of analytic functions with bounded mean modulus*, Trudy Mat. Inst. Steklov. **18** (1946). (Russian)

22. ______, *On the Carathéodory-Fejér problem and a similar problem*, Mat. Sb. **18 (60)** (1946), 213–226; reprinted as Chapter XI, §7 in his *Geometric theory of functions of a complex variable*, 2nd ed., "Nauka", Moscow, 1966, pp. 477–492; English transl., Amer. Math. Soc., Providence, R. I., 1969, pp. 497–514.

23. E. G. Gol'shteĭn, *Duality theory in mathematical programming and its applications*, "Nauka", Moscow, 1971; German transl., *Dualitätstheorie in der nichtlinearen Optimierung und ihre Anwendung*, Akademie-Verlag, Berlin, 1975.

24. Helmut Grunsky, *Eindeutige beschränkte Funktionen in mehrfach zusammen-Langenden Gebieten*. I, II, Iber. Deutsch. Math. Verein. **50** (1940), 230–255; **52** (1942), 118–132.

25. Nelson Dunford and Jacob T. Schwartz, *Linear operators*. Vol. I: *General theory*, Interscience, 1958.

26. J. L. Doob, *A minimum problem in the theory of analytic functions*, Duke Math. J. **8** (1941), 413–424.

27. Peter L. Duren, *Theory of H^p spaces*, Academic Press, 1970.

28. Lennart Carleson and Sigvard Jacobs, *Best uniform approximation by analytic functions*, Ark. Mat. **10** (1972), 219–229.

29. James G. Caughran, *Nonuniqueness of extremal kernels*, Proc. Amer. Math. Soc. **18** (1967), 957–958.

30. A. N. Kolmogorov and S. V. Fomin, *Elements of the theory of functions and of functional analysis*, 2nd ed., "Nauka", Moscow, 1968; English transl. of 1st ed., vols. I, II, Graylock Press, Albany, N. Y., 1957, 1961.

31. N. P. Korneĭchuk [M. P. Korniĭchuk], *Extremal problems of approximation theory*, "Nauka", Moscow, 1976. (Russian)

32. A. N. Kochetkov, *Extremal problems with asymmetric supplementary condition in certain classes of analytic functions*, Akad. Nauk Armyan. SSR Dokl. **41** (1965), 135–139. (Russian)

33. M. G. Kreĭn, *The L-problem in an abstract linear normed space*, Article 4 in N. I. Akhiezer and M. G. Kreĭn, *Some questions in the theory of moments*, Gos. Nauchno-Takhn. Izdat. Ukraïn., Kharkov, 1937, pp. 171–199, English transl., Amer. Math. Soc., Providence, R. I., 1962, pp. 175–204.

34. V. M. Adamyan, D. Z. Arov and M. G. Kreĭn, *Infinite Hankel matrices and the generalized problems of Carathéodory-Fejér and F. Riesz*, Funktsional. Anal. i Prilozhen. **2** (1968), no. 1, 1–19; English transl. in Functional Anal. Appl. **2** (1968).

35. ______, *Infinite Hankel matrices and the generalized problems of Carathéodory-Fejér and Schur*, Funktsional. Anal. i Prilozhen. **2** (1968), no. 4, 1–17; English transl. in Functional Anal. Appl. **2** (1968).

36. M. G. Kreĭn and A. A. Nudel'man, *The Markov moment problem and extremal problems*, "Nauka", Moscow, 1973; English transl., Amer. Math. Soc., Providence, R. I., 1977.

37. Edmund Landau, *Abschätzung der Koeffizientensumme einer Potenzreihe*. I, II, Arch. Math. und Phys. (3) **21** (1913), 42–50, 250–255.

38. Pierre-Jean Laurent, *Approximation et optimisation*, Hermann, Paris, 1972.

39. Karel de Leeuw and Walter Rudin, *Extreme points and extremum problems in H_1*, Pacific J. Math. **8** (1958), 467–485.

40. Peter D. Lax, *Reciprocal extremal problems in function theory*, Comm. Pure Appl. Math. **8** (1955), 437–453.

41. A. J. Macintyre and W. W. Rogosinski, *Extremum problems in the theory of analytic functions*, Acta Math. **82** (1950), 275–325.

42. G. M. Mordasova, *Estimate of the remainder in a Taylor series for analytic functions with bounded rth derivative*, Uspekhi Mat. Nauk **14** (1959), no. 4 (88), 187–194. (Russian)

43. Zeev Nehari, *Bounded analytic functions*, Bull. Amer. Math. Soc. **57** (1951), 354–366.

44. S.M. Nikol'skiĭ, *Approximation of functions in the mean by trigonometric polynomials*, Izv. Akad Nauk SSSR Ser. Mat. **10** (1946), 207–256; English transl. in Amer. Math. Soc. Transl. (2) **102** (1973).

45. K. Yu. Osipenko, *Optimal interpolation of analytic functions*, Mat. Zametki **12** (1972), 465–476; English transl. in Math. Notes **12** (1972).

46. Jacqueline Penez, *Approximation by boundary values of analytic functions*, Proc. Nat. Acad Sci. U. S. A. **40** (1954), 240–243.

47. G. Pólya and G. Szegö, *Aufgaben und Lehrsatz aus der Analysis*. Vols. I, II, 2nd ed., Springer-Verlag, 1954; rev. English transl., 1972, 1975.

48. Arthur H. Read, *A converse of Cauchy's theorem and applications to extremal problems*, Acta Math. **100** (1958), 1–22.

49. ______, *Conjugate extremal problems of class* $p = 1$, Ann. Acad. Sci. Fenn. Ser. A I Math. No. 250/28 (1958).

50. Friedrich [Frigyes] Riesz, *Über Potenzreihen mit vorgeschriebenen Anfangsgliedern*, Acta Math. **42** (1918/20), 145–171.

51. Raphael M. Robinson, *Analytic functions in circular rings*, Duke Math. J. **10** (1943), 341–354.

52. W. W. Rogosinski and H. S. Shapiro, *On certain extremum problems for analytic functions*, Acta Math. **90** (1953), 287–318.

53. H. L. Royden, *The boundary values of analytic and harmonic functions*, Math. Z. **78** (1962), 1–24.

54. Walter Rudin, *Analytic functions of class* H_p, Trans. Amer. Math. Soc. **78** (1955), 46–66.

55. Ottó Szász, *Ungleichheitsbeziehungen für die Ableitungen einer Potenzreihe, die eine im Einheitskreise beschränkte Funktion darstellt*, Math. Z. **8** (1920), 303–309.

56. ______, a) *Über die Koeffizienten beschränkter Potenzreihen*, Mat. és Természet. Értesitö **43** (1926), 488–503. (Hungarian; German summary) b) *Über beschränkte Potenzreihen*, Mat. és Természet. Értesitö **43** (1926), 504–520. (Hungarian; German summary)

57. S. B. Stechkin, *Estimates of the remainder in the Taylor series for certain classes of analytic functions*, Izv. Akad. Nauk SSSR Ser. Mat. **17** (1953), 461–472. (Russian)

58. V. M. Tarpigoreva, *Extremal problems for Orlicz classes of analytic functions in the unit circle*, Litovsk. Mat. Sb. **3** (1963), no. 1, 237–269. (Russian)

59. V. M. Tikhomirov, *Some questions in approximation theory*, Izdat. Moskov. Gos. Univ., Moscow, 1976. (Russian)

60. G. Ts. Tumarkin and S. Ya. Khavinson, *Investigation of the properties of extremal functions by means of duality relations in extremal problems for classes of analytic functions in multiply connected domains*, Mat. Sb. **46** (88) (1958), 195–228. (Russian)

61. ______, *Qualitative properties of solutions of certain types of extremal problems*, Studies of Current Problems of the Theory of Functions of a Complex Variable (Third All-Union Conf., Moscow, 1957; A. I. Markushevich, editor), Fizmatgiz, Moscow, 1960, pp. 77–95; French transl., Gauthier-Villars, Paris, 1962, pp. 73–92.

62. J. L. Walsh, *The location of critical points of analytic and harmonic functions*, Amer. Math. Soc., Providence, R. I., 1950.

63. S. Ya. Khavinson, *On an extremal problem of the theory of analytic functions*, Uspekhi Mat. Nauk **4** (1949), no. 4 (32), 158–159. (Russian)

64. ______, *On some extremal problems of the theory of analytic functions*, Moskov. Gos. Univ. Uchen. Zap. **148** (1951) (Mat. 4), 133–143; English transl. in Amer. Math. Soc. Transl. (2) **32** (1963).

65. ______, *On extremal properties of functions mapping a domain onto a multi-sheeted disc*, Dokl. Akad. Nauk SSSR **88** (1953), 957–959. (Russian)

66. ______, *Extremal problems for certain classes of analytic functions in finitely connected domains*, Mat. Sb. **36 (78)** (1955), 445–478; English transl. in Amer. Math. Soc. Transl. (2) **5** (1957).

67. ______, *The analytic capacity of plane-sets, some classes of analytic functions, and the extremum function in Schwarz's lemma for arbitrary domains*, Dokl. Akad. Nauk SSSR **128** (1959), 896–898. (Russian)

68. ______, *The theory of extremal problems for bounded analytic functions satisfying additional conditions inside the domain*, Uspekhi Mat. Nauk **18** (1963), no. 2 (110), 25–98; English transl. in Russian Math. Surveys **18** (1963).

69. ______, *Analytic functions of bounded type*, Itogi Nauki: Mat. Anal. 1963, VINITI, Moscow, 1965, pp. 5–80. (Russian)

70. ______, *An estimate of the Taylor sums of bounded analytic functions in the disk*, Dokl. Akad. Nauk SSSR **80** (1951), 333–336. (Russian)

71. ______, *Representation of extremal functions in the classes* E_q *in terms of Green's and Neumann's functions*, Mat. Zametki **16** (1974), 707–716; English transl. in Math. Notes **16** (1974).

72. Maurice H. Heins, *Extremal problems for functions analytic and single-valued in a doubly-connected region*, Amer. J. Math. **62** (1940), 91–106.

73. Dennis A. Hejhal, *Sur les problèmes linéaires extrémaux pour les fonctions analytiques*, C. R. Acad. Sci. Paris Sér. A–B **275** (1972), A329–A332.

74. ______, *Classification theory for Hardy classes of analytic functions*, Ann. Acad. Sci. Fenn. Ser. A I Math. No. 566 (1973).

75. Menahem Schiffer and Donald C. Spencer, *Functionals of finite Riemann surfaces*, Princeton Univ. Press, Princeton, N. J., 1954.

APPENDIX

D. KHAVINSON

Smirnov Classes of Analytic Functions in Multiply Connected Domains

Let G be an n-connected domain in $\mathbf{C}$ bounded by the simple rectifiable contours $\gamma_1, \ldots, \gamma_n$, $\Gamma = \bigcup_1^n \gamma_j$. As usual, $g(\zeta, z)$ denotes the Green function of G, i.e. $g(\zeta, z) = -\ln|\zeta - z| + u_\zeta(z)$, where $u_\zeta(z)$ is harmonic in G as a function of z and $g(\zeta, z)|_\Gamma \equiv 0$ for all $z \in G$. The measure

$$d\omega_z = \frac{1}{2\pi} \cdot \frac{\partial g(\zeta, z)}{\partial n_\zeta}\, ds$$

on Γ is the harmonic measure at the point $z \in G$. Here n_ζ is the inner normal towards Γ at ζ and $ds = d|\zeta|$ is Lebesgue measure on Γ.

Fix $z_0 \subset G$. As is known, the Radon-Nikodým derivative

$$K(\zeta, z) = \frac{1}{2\pi} \frac{d\omega_z(\zeta)}{d\omega_{z_0}(\zeta)}, \qquad \zeta \in \Gamma,\ z \in G,$$

plays the role of the classical Poisson kernel in G. Since G can be mapped conformally onto a domain bounded by simple analytic contours and the harmonic measure is invariant under the conformal mappings, it is readily seen that $K(\zeta, z)$ is harmonic in G for all $\zeta \in \Gamma$ and continuous on Γ for all $z \in G$. Moreover, if $u(z)$ is a harmonic function in G such that the subharmonic function $|u|$ has a harmonic majorant in G, then there exists a Borel measure μ on Γ such that

$$u(z) = \int_\Gamma K(z, \zeta)\, d\mu(\zeta). \tag{1}$$

For further details, we refer the reader to [2], Chapters 1–4, or [5], §1.([1])

A function f analytic in G is said to belong to the Nevanlinna class N or to the Hardy class H^p, $p > 0$, if the subharmonic function $\ln^+|f|$ or, respectively, $|f|^p$ has a harmonic majorant in G.

It is known that, as in the classical case of the unit disk (see [1], Chapter 10; [2], Chapter 6, or [9]), $f \in N \Leftrightarrow f = \phi/\psi$, where ϕ and ψ are bounded analytic functions in G with $\psi \neq 0$ in G.

([1])These numbers refer to the separate bibliography at the end of this Appendix.

If $f \in N$, and $\{z_n\}_1^\infty$ are the zeros of f in G, then $\Sigma_1^\infty g(z, z_j)$ converges uniformly on compact subsets of $G \setminus \{z_n\}_1^\infty$. This is the generalization of the well-known Blaschke condition in the unit disk. (See [2], §4.7. Also see [5], §3, for further discussion.)

A function f analytic in G is said to belong to the class D (or N_+) if there exists a sequence of n-connected domains $\{G^i\}$ such that $z_0 \in G^1 \subset G^2 \subset \cdots$, $\bigcup_1^\infty G^i = G$ and for each $\varepsilon > 0$ there is a $\delta > 0$ such that $E \subset \partial G^i$ and $\omega_{z_0}^i(E) < \delta$ imply that

$$\int_E \ln^+|f|\, d\omega_{z_0}^i < \varepsilon \tag{2}$$

for all i. Here $d\omega_{z_0}^i$ is the harmonic measure on ∂G^i taken at z_0. Condition (2) is called the *uniform integrability* of $\ln^+|f|$ with respect to the harmonic measure. As is seen from the definitions, the classes N, D and all the H_p, $p > 0$, are conformally invariant. Therefore, it suffices to study those classes in domains with analytic boundaries. It is not hard to see that $N \supset D$, $N \supset H_p$ for all p, and $H_{p_1} \supset H_{p_2}$ for $p_1 \leqslant p_2$. The inclusion $D \supset H_p$ for all $p > 0$, which is known to be true in the unit disk, also holds in this case (see [1], §2.5, [9], or [5]). Moreover, similarly to the case of the unit disk, one can show that $f \in H_p$, $p \geqslant 1$, if and only if

$$f(z) = \int_\Gamma K(\zeta, z) f(\zeta)\, d\omega_{z_0}, \tag{3}$$

where $f(\zeta) \in L_p(d\omega_{z_0})$ represents the angular boundary values of f almost everywhere (with respect to $d\omega_{z_0}$) on Γ. (More details can be found in [1], Chapter 10, [2], Chapter 6, and [9].) From (3) it follows that $\|f\|_p = \|f(\zeta)\|_{L_p(d\omega_{z_0})}$ defines a norm on H_p for $p \geqslant 1$. Equipped with this norm, H_p, $p \geqslant 1$, becomes a Banach space.

If $f \in N$ and $\{z_n\}_1^\infty$ are the zeros of f, then the harmonic function $\ln|f| + \Sigma_1^\infty g(z, z_k)$ is representable by the integral (1) with $d\mu = \ln|f(\zeta)|\, d\omega_{z_0} + d\nu$, where $f(\zeta)$ represents boundary values of f and $d\nu$ is singular. Moreover, one can show that $f \in D$ if and only if $d\nu \leqslant 0$ (cf. [2], Chapter 6, [9] and [5]).

A function $f(z)$ analytic in G is said to belong to the Smirnov class E_p, $p > 0$, if there exists a sequence of domains $\{G^i\}$, $G^1 \subset G^2 \subset \cdots$, $\bigcup_i G^i = G$, with rectifiable boundaries and a constant $M > 0$ such that

$$\int_{\partial G^i} |f(z)|^p\, ds \leqslant M < +\infty.$$

As is seen from the definition, the classes E_p are not conformally invariant. Their properties strongly depend on the "smoothness" of the boundary Γ of G–more precisely, on the relation between the harmonic measure and Lebesgue measure on Γ. The following theorem, characterizing the connection between E_p and H_p classes, was proved by M. V. Keldysh and M. A. Lavrent'ev for simply connected domains, and by Tumarkin and Khavinson for finitely connected domains (see [1], Chapter 10, [7] and [9]).

THEOREM 1. *Let* $\phi\colon G \to K$ *be a conformal mapping of* G *onto a domain* K *bounded by simple analytic contours. Then* $f(z) \in E_p(G)$ *if and only if* $f(z)/\sqrt[p]{\phi'(z)} \in H_p(G)$.

From Theorem 1 it follows that if $f \in E_p$, then $f \in N$ and, therefore, f has angular boundary values $f(\zeta)$ a.e. on Γ. Moreover, one can show that $f(\zeta) \in L_p(ds)$, and that $\|f\|_{E_p} = \|f(\zeta)\|_{L_p(\Gamma, ds)}$ defines a norm on E_p, $p \geqslant 1$. The space $E_p(G)$, $p \geqslant 1$, equipped with this norm, becomes a Banach space. (For more details and further references see [1], Chapter 10, [7] and [9].)

There is a special class of domains where properties of the E_p-classes are quite similar to those of the H_p-classes.

Let $\phi\colon G \to K$ be a conformal mapping of G onto the domain K with an analytic boundary. In view of the celebrated theorem of Köebe, we can always assume that K is bounded by circles (see [3], Chapter 5, §6). Then it easily follows from Theorem 1 and the rectifiability of Γ that $\phi'(z) \in E_1(G)$, and therefore $(E_1(G) \in N)$

$$\ln|\phi'(z)| = \int_\Gamma K(\zeta, z)\left[\ln|\phi'(\zeta)|\,d\omega_{z_0} + d\nu_G\right], \tag{4}$$

where $d\nu_G$ is singular. Since $1/\phi'(z) \in H_1(G)$ by Theorem 1, and hence $1/\phi'(z) \in D$, we obtain from (4) that $d\nu_G \geqslant 0$.

DEFINITION. A domain G is called a *Smirnov domain* $(G \in S)$ if $d\nu_G \equiv 0$.

A quite detailed survey on simply connected Smirnov and non-Smirnov domains, corresponding properties of E_p-classes and related historical remarks can be found in [1], Chapter 10, and [3], Chapter 10. "Abnormal" behaviour of E_p-functions in non-Smirnov domains is discussed in [4]. The definition of the class S for multiply connected domains is due to Tumarkin and Khavinson. We refer the reader to their articles [7]–[9] for greater detail and further references. Also, see Tumarkin's papers [10] and [11].

The following important generalization of the Smirnov theorem (Theorem 2.11 in [1]) is due to Tumarkin and Khavinson (see [8] and [9]; also, see [4]).

THEOREM 2. *Let* $G \in S$. *Then* $f(z) \in E_p(G)$ *if and only if* $f \in D$ *and* $f(\zeta) \in L_p(\Gamma, ds)$.

As we mentioned above, $E_p(G) \subset N$ for all $p > 0$. Also, it is clear that $E_{p_1}(G) \supseteq E_{p_2}(G)$ for $p_1 \leqslant p_2$.

If $G \notin S$, then $E_p \not\subset D$ for any $p > 0$. The following characterization of E_p-classes in general domains has been obtained in [4] (cf. Theorem 4.6).

A function $f \in N$ analytic in G with zeros $\{z_n\}_1^\infty$ belongs to $E_p(G)$ if and only if

$$\ln|f(z)| + \sum_1^\infty g(z, z_n) = \int_\Gamma K(\zeta, z)\left[\ln|f(\zeta)|\,d\omega_{z_0} + d\nu_f\right], \tag{5}$$

where $f(\zeta) \in L_p(\Gamma, ds)$, $d\nu_f \perp ds$ and $d\nu_f \leqslant (1/p)\, d\nu_G$. In particular, (5) implies that $\bigcap_{0<p<\infty} E_p(G) \subset D$ for any G (Corollary 4.2 in [4]).

We finish the discussion on E_p-classes in Smirnov domains with the following important result of Tumarkin generalizing the classical result of V. I. Smirnov for $n = 1$, $p = 2$ (cf. Theorem 10.7 in [1] and §7 in [3]).

If $G \in S$, then $E_p(G)|_\Gamma$ is the $L_p(ds, \Gamma)$-closure of the rational functions with poles outside of $\overline{G}$.

The following generalization of the classical F. and M. Riesz theorem has been discovered independently by many authors. (For further references see Theorem 10.4 in [1], Theorem 4.4 in [2], Chapter 10, §5 in [3], and [9].)

THEOREM 3. *Let $\Gamma = \partial G$ be rectifiable, and let $d\mu(\zeta)$ be a complex Borel measure on Γ. Then*

$$\frac{1}{2\pi i}\int_\Gamma \frac{d\mu(\zeta)}{\zeta - z} \equiv 0, \qquad z \in C\setminus G,$$

if and only if $d\mu(\zeta) = f(\zeta)\, d\zeta$, where $f(\zeta)$ represents the boundary values of a function $f(z)$ of the class $E_1(G)$.

An obvious difference between the H_p- and E_p-classes, i.e. the difference between the harmonic measure and Lebesgue measure, poses a natural question: When do those classes coincide? This is answered by the following theorem proved by Keldysh and Lavrent'ev for simply connected domains and by Tumarkin and Khavinson for multiply connected domains (cf. [1], Theorem 10.2, and [9]).

Let $\phi\colon G \to K$ be as above. Then $H_p(G) = E_p(G)$ if and only if

$$0 < \text{const}_1 \leqslant |\phi'(z)| \leqslant \text{const}_2 < +\infty.$$

The properties of E_p-spaces with respect to the weak topologies can be combined into the following theorem due to Khavinson (see [6], Theorem 6.3; also see Corollary 4.2 in [2]).

THEOREM 4. *Let G be an n-connected domain with rectifiable boundary Γ.*

(i) *If the sequence of analytic functions $f_n(z)$ is uniformly bounded in G and converges to the function $f(z)$, uniformly on compact subset of G, then $f(z)$ is bounded in G and, for an arbitrary function $\omega(\zeta) \in L_1(\Gamma, ds)$,*

$$\lim_{n\to\infty} \int_\Gamma f_n(\zeta)\omega(\zeta)\, d\zeta = \int_\Gamma f(\zeta)\omega(\zeta)\, d\zeta. \tag{6}$$

(ii) *If the sequence of E_p-functions $\{f_n(z)\}$, $1 < p < +\infty$, converges to $f(z)$ uniformly on compact subsets of G and the E_p-norms of the f_n are uniformly bounded, then $f(z) \in E_p(G)$ and (6) holds for all $\omega(\zeta) \in L_q(\Gamma, ds)$, where $1/p + 1/q = 1$.*

(iii) *If the sequence of E_1-functions $\{f_n(z)\}$ converges to $f(z)$ uniformly on compact subsets of G and the E_1-norms of the f_n are uniformly bounded, then $f(z) \in E_1(G)$ and (6) holds for all $\omega(\zeta) \in C(\Gamma)$.*

NOTE. In (iii) one cannot replace $C(\Gamma)$ by $L_\infty(\Gamma)$.

REFERENCES

1. Peter L. Duren, *Theory of H^p spaces*, Academic Press, 1970.

2. Stephen D. Fisher, *Function theory on planar domains*, Wiley, 1983.

3. G. M. Goluzin, *Geometric theory of functions of a complex variable*, 2nd rev. ed., "Nauka", Moscow, 1966; English transl., Amer. Math. Soc., Providence, R. I., 1969.

4. D. Khavinson, *Remarks concerning boundary properties of analytic functions of E_p-classes*, Indiana Univ. Math. J. **31** (1982), 779–787.

5. ______, *Factorization theorems for different classes of analytic functions in multiply connected domains*, Pacific J. Math. **108** (1983), 295–318.

6. S. Ya. Khavinson, *Extremal problems for certain classes of analytic functions in finitely connected domains*, Mat. Sb. **36 (78)** (1955), 445–478; English transl. in Amer. Math. Soc. Transl. (2) **5** (1957).

7. G. Ts. Tumarkin and S. Ya. Khavinson, *On the definition of analytic functions of class E_p in multiply connected domains*, Uspekhi Mat. Nauk **13** (1958), no. 1 (79), 201–206. (Russian)

8. ______, *Analytic functions in multiply connected domains of Smirnov type* (*class S*), Izv. Akad. Nauk SSSR Ser.Mat. **22** (1958), 379–386. (Russian)

9. ______, *Classes of analytic functions in multiply connected domains*, Studies of Current Problems of the Theory of Functions of a Complex Variable (Third All-Union Conf., Moscow, 1957; A. I. Markushevich, editor), Fizmatgiz, Moscow, 1960, pp. 45–77; French transl., Gauthier-Villars, Paris, 1962, pp. 37–71.

10. G. Ts. Tumarkin, *A sufficient condition for a domain to belong to class S*, Vestnik Leningrad. Univ. **1962**, no. 13 (Ser. Mat. Mekh. Astr. vyp. 3), 47–55. (Russian)

11. ______, *Effect of a deformation of the boundary on certain conformal properties of domains* (*membership of the domain in V. I. Smirnov's class S*), Mat. Zametki **8** (1970), 351–359; English transl. in Math. Notes **8** (1970).

Amer. Math. Soc. Transl.
(2) Vol. **129**, 1986

Foundations of the Theory of Extremal Problems for Bounded Analytic Functions with Additional Conditions*

S. YA. KHAVINSON

Preface

This text is a continuation of [55], and uses the same notation. It is addressed to students of the mathematical department of the faculty for continuing education of instructors at the Kuibyshev Institute of Civil Engineering in Moscow. We hope that it will be useful first of all for those students who are undertaking to master function theory, functional analysis, and the theory of extremal problems. Along with the classical theorems of Carathéodory-Fejér, Milloux, and others, we also present here some recent research: the duality method for analytic functions with supplementary restrictions on growth inside the domain, a general approach to extremal problems in the classes of nonvanishing functions in H_p (Hardy classes), and some other topics.

The historical notes at the end of the paper and the list of references do not claim to be complete.

I take this opportunity to express my gratitude to my colleagues at the Department of Higher Mathematics in our Institute, above all to N. G. Belousov, for their help in preparation of the manuscript.

CHAPTER I. THE CARATHÉODORY-FEJÉR PROBLEM

§1. Generalized Carathéodory-Fejér problem

In 1911, Carathéodory and Fejér [27] studied the following extremal problem for analytic functions in the unit disk. Among functions whose expansion in the neighborhood of $z = 0$ starts with $c_0 + c_1 z + \cdots + c_{m-1} z^{m-1}$, where

1980 *Mathematics Subject Classification* (1985 *Revision*). Primary 30D55; Secondary 30E99.

*Translation of a set of lecture notes published as a textbook (for teachers' continuing education), Moskov. Inzh.-Stroit. Inst., Moscow, 1981.

$c_0, c_1, \ldots, c_{m-1}$ are given numbers, find a function which has the smallest maximum modulus in the disk $|z| \leqslant 1$. This problem is closely related to the description of the set of Taylor coefficients $(c_0, c_1, \ldots, c_{m-1})$ of functions in the class B^1, especially of the functions corresponding to the boundary points of that set. Those problems have been generalized in many directions. In this section we consider a natural generalization of such problems to the case of an arbitrary Banach space and then apply our results to the space $B(G)$ of functions bounded in a given finitely connected domain G.

Let $X = \{f\}$ be a (complex) Banach space with a unit sphere $X^1 = S$. Let $l_1(f), \ldots, l_m(f)$ be continuous, linearly independent linear functionals on X. Assume that the unit sphere S is sequentially compact in the topology determined by the functionals $l_1, \ldots, l_m$. This means that from any sequence $\{f_n\} \subset S$ we can choose a subsequence $\{f_{n_k}\}$ such that there is an $f \in S$ for which

$$\lim_{k \to \infty} l_i(f_{n_k}) = l_i(f), \qquad i = 1, \ldots, m. \tag{1.1}$$

Let $\mathbf{C}^m$ be an m-dimensional complex Euclidean space consisting of points $c = (c_1, \ldots, c_m)$. Setting

$$c_i = l_i(f), \qquad i = 1, \ldots, m, \tag{1.2}$$

we obtain a mapping of X into $\mathbf{C}^m$. Let A_m denote the image of S under this mapping. Finally, let L denote the linear span of $l_1, \ldots, l_m$ in the dual space X^* of X.

THEOREM 1.1. *A_m is a convex bounded closed set, and $(0, 0, \ldots, 0)$ is an interior point of A_m. If $c^* = (c_1^*, \ldots, c_m^*)$ is an arbitrary boundary point of A_m, then c^* is the image of an element $f^* \in S$, $\|f^*\| = 1$, which is extremal for some functional $l \in L$, $l \neq 0$. Conversely, if an element f^*, $\|f^*\| = 1$, is extremal for a functional $l \in L$, $l \neq 0$, then the point $(l_1(f^*), \ldots, l_m(f^*))$ is a boundary point of A_m.*

(An element $\mathcal{F} \neq 0$ is said to be *extremal for the functional l* if $l(\mathcal{F}) = \|l\| \cdot \|\mathcal{F}\|$.)

PROOF. It is obvious that A_m is convex and bounded. That it is also closed follows from (1.1). When f runs over the space X, the point c defined by (1.2) describes a certain subspace A in $\mathbf{C}^m$. If $A \neq \mathbf{C}^m$, then A lies entirely in a hyperplane in $\mathbf{C}^m$ passing through the origin. This implies that

$$\lambda_1 l_1(f) + \cdots + \lambda_m l_m(f) = 0 \qquad \forall f \in X \tag{1.3}$$

with constants $\lambda_1, \ldots, \lambda_m$ not all equal to zero. But (1.3) contradicts the linear independence of $l_1, \ldots, l_m$. Thus, $A = \mathbf{C}^m$. In particular, this implies that $0 = (0, \ldots, 0)$ is an inner point of A_m.

Let $c^* = (c_1^*, \ldots, c_m^*)$, $c_k^* = x_k^* + iy_k^*$, $k = 1, \ldots, m$, be a boundary point of A_m. Let us treat $\mathbf{C}^m$ as a real $2m$-dimensional Euclidean space $R^{2m} = \{(x_1, y_1, \ldots, x_m, y_m)\}$, $c_k = x_k + iy_k$, $k = 1, \ldots, m$. Since A_m is a convex body in R^{2m}, there exists a supporting hyperplane

$$\sum_{k=1}^{m} \alpha_k x_k + \beta_k y_k = d \qquad (d > 0) \tag{1.4}$$

passing through c^* and such that for all points $c = (x_1, y_1, \ldots, x_m, y_m) \in A_m$

$$\sum_{k=1}^{m} \alpha_k x_k + \beta_k y_k \leqslant d. \tag{1.5}$$

This becomes an equality for the point c^*. Set $\gamma_k = \alpha_k - i\beta_k$. Then, for all points $c \in A_m$, we can rewrite (1.5) as follows:

$$\operatorname{Re} \sum_{k=1}^{m} \gamma_k c_k \leqslant d. \tag{1.6}$$

But since together with a point c A_m contains also all points $e^{i\alpha}c$ for an arbitrary real α, for each $c \in A_m$ we have

$$\left| \sum_{k=1}^{m} \gamma_k c_k \right| \leqslant d. \tag{1.7}$$

Again, (1.7) becomes an equality for c^*.

Consider a linear functional $l \in L$,

$$l = \sum_{k=1}^{m} \gamma_k l_k. \tag{1.8}$$

Then

$$\|l\| = \sup_{f \in S} \left| \sum_{k=1}^{m} \gamma_k l_k(f) \right| = \sup \left| \sum_{k=1}^{m} \gamma_k c_k \right| = d. \tag{1.9}$$

Let $f^* \in S$ be a preimage of c^* under the mapping (1.2). Then

$$|l(f^*)| = \left| \sum_{k=1}^{m} \gamma_k c_k^* \right| = d. \tag{1.10}$$

On the other hand,

$$|l(f^*)| \leqslant \|l\| \cdot \|f^*\| \leqslant d\|f^*\|. \tag{1.11}$$

Therefore, $\|f^*\| = 1$ and in (1.11) we have equality. So, f^* is extremal for l.

Conversely, let f^*, $\|f^*\| = 1$, be extremal for a functional l of the form (1.8). We can assume that

$$l(f^*) = \|l\|. \tag{1.12}$$

Then

$$|l(f)| = \left| \sum_{k=1}^{m} \gamma_k l_k(f) \right| = \left| \sum_{k=1}^{m} \gamma_k c_k \right| \leqslant \|l\| \qquad \forall c \in A_m,$$

and a fortiori

$$\operatorname{Re} \sum_{k=1}^{m} \gamma_k c_k \leqslant \|l\| \qquad \forall c \in A_m. \tag{1.13}$$

At the same time, according to (1.12),

$$\operatorname{Re}\sum_{k=1}^{m}\gamma_k c_k^* = \operatorname{Re}\sum_{k=1}^{m}\gamma_k l_k(f^*) = \operatorname{Re} l(f^*) = \|l\|. \tag{1.14}$$

From (1.13) and (1.14) it follows that $\operatorname{Re}\sum_1^m \gamma_k c_k = \|l\|$ is a supporting hyperplane of A_m and the point c^* belongs to it; hence c^* is a boundary point of A_m.

THEOREM 1.2. *Let $c_1, \dots, c_m$ be arbitrary numbers (not all equal to zero). Among the elements of X satisfying the interpolation conditions*

$$l_i(f) = c_i, \qquad i = 1, \dots, m, \tag{1.15}$$

there exists an element of minimal norm. Every such element f^* of minimal norm is extremal for a certain (common for all such elements) linear functional $l \in L$ and, therefore, has the form

$$f^* = \lambda f^{**}, \tag{1.16}$$

where f^{**} is the element of unit norm corresponding to a boundary point of A_m.

PROOF. Since the origin is an interior point of A_m, for sufficiently small $\sigma > 0$ the point $\sigma c = (\sigma c_1, \dots, \sigma c_m) \in A_m$. Let σ^* be the least upper bound of all σ such that $\sigma c \in A_m$. Since A_m is bounded, $\sigma^* < +\infty$. Since A_m is closed, $\sigma^* c \in A_m$. Moreover, clearly $\sigma^* c$ is a boundary point of A_m. Otherwise there would exist $\sigma > \sigma^*$ such that $\sigma c \in A_m$. According to Theorem 1.1, there exists an element f^{**}, $\|f^{**}\| = 1$, extremal for a certain functional $l \in L$ and such that $l_i(f^{**}) = \sigma^* c_i$, $i = 1, \dots, m$. Then the element $f^* = (\sigma^*)^{-1} f^{**}$ satisfies (1.15) and also

$$|l(f^*)| = \|l\| \cdot \|f^*\|. \tag{1.17}$$

Since $l = \sum_1^m \gamma_k l_k$, for each element φ satisfying (1.15) we have

$$|l(f^*)| = |l(\varphi)| \leqslant \|l\| \cdot \|\varphi\|. \tag{1.18}$$

Hence $\|f^*\| \leqslant \|\varphi\|$. If $\|f^*\| = \|\varphi\|$ for some φ satisfying (1.15), then (1.18) implies that φ is also an extremal element for the functional l.

THEOREM 1.3. *If for each functional $l \in L$, $l \neq 0$, there exists at most one extremal element (up to a constant factor), then to each boundary point of A_m corresponds a unique element of S. Also, among elements satisfying* (1.15) *there is a unique element of minimal norm.*

PROOF. Every element corresponding to a boundary point of A_m must be extremal for some functional $l \in L$ determined by this point. Therefore, if f_1 and f_2 are two elements corresponding to the same boundary point, they can only differ by a constant factor t. From the equalities $l_i(f_1) = l_i(f_2)$, $i = 1, \dots, m$, it follows that $t = 1$. Similarly one can prove the second assertion of the theorem.

Let G be an n-connected domain bounded by rectifiable contours $\gamma_1, \dots, \gamma_n$, $\Gamma = \bigcup_1^n \gamma_j$. Let functions $\omega_1(\zeta), \dots, \omega_m(\zeta)$ be analytic on Γ, while none of their nontrivial linear combinations are analytic in the whole domain G. Let us consider the linear functionals

$$l_k(f) = \int_\Gamma f(\zeta)\omega_k(\zeta)\,d\zeta, \qquad k = 1, \dots, m, \tag{1.19}$$

in the space $B(G)$ of bounded analytic functions. By virtue of the condition imposed on $\{\omega_k(\zeta)\}_1^m$ these functionals are linearly independent. Consider the class $B^1_{\rho(\zeta)}$ as the unit sphere in $B(G)$. This means that we have introduced the following metric in $B(G)$:

$$\|f\| = \operatorname*{ess\,sup}_{\zeta\in\Gamma} \left|\frac{f(\zeta)}{\rho(\zeta)}\right|. \tag{1.20}$$

The function $\rho(\zeta) > 0$ is assumed to be continuous.

Combining the results of §3 and Theorem 6.2 in [55] with the previous theorems in this section, we obtain the following result.

THEOREM 1.4. *When $f(\zeta)$ runs over the class B^1_ρ, the points $(l_1(f), \dots, l_m(f))$ fill out a bounded closed convex set A_m in $\mathbf{C}^m$ containing the origin as an interior point. To each boundary point of A_m corresponds a unique function $f^*(z) \in B^1_\rho$ which has the form*

$$f^*(z) = e^{i\delta}\exp\left[\frac{1}{2\pi}\int_\Gamma \ln\rho(\zeta)\frac{\partial P(z,\zeta)}{\partial n}\,ds - \sum_{j=1}^k P(z,\alpha_j)\right], \tag{1.21}$$

where $P(z,\zeta)$ is the complex Green function, $\alpha_1, \dots, \alpha_k$ are certain points (zeros of f^*), and δ is a real constant (see (6.1) *in* [55]). *Given arbitrary numbers $c_1, \dots, c_m$ not all equal to zero, among the functions of the space $B(G)$ satisfying the interpolation conditions* (1.15) *there exists a unique function for which*

$$\inf \operatorname*{ess\,sup}_{\zeta\in\Gamma} \left|\frac{f(\zeta)}{\rho(\zeta)}\right| \tag{1.22}$$

is attained. This function $f^*(z)$ has the form $f^*(z) = \lambda f^{**}(z)$, where $\lambda > 0$ and f^{**} is again defined by (1.21).

For future reference we note that when the domain G is the unit disk and $\rho(\zeta) \equiv 1$, (1.21) takes the following form:

$$f^*(z) = e^{i\delta}\prod_{j=1}^k \frac{z-\alpha_j}{1-\bar{\alpha}_j z} \tag{1.21'}$$

and, therefore, $f^*(z)$ is a finite Blaschke product.

Since $\rho(\zeta) = 1$, by using Theorem 7.1 in [55] we obtain the following result.

THEOREM 1.5. *The image of the class $B^1(G)$ in $\mathbf{C}^m$ under the mapping* (1.19) *is a bounded closed convex set A_m containing the origin as an interior point. To each boundary point of A_m corresponds a unique function $f^*(z) \in B^1$. Moreover, either $f^*(z) \equiv e^{i\delta} \equiv$ const or $f^*(z)$ maps G onto the multi-sheeted unit disk with at least* *n sheets. Given arbitrary numbers* $c_1, \dots, c_n$ *not all equal to zero, the unique function among those of the class* $B(G)$ *satisfying interpolation conditions* (1.15) *for which the infimum*

$$\inf \sup_{z \in G} |f(z)| \tag{1.23}$$

is attained has the form $f^*(z) = \lambda f^{**}(z)$. *Here again either* $f^*(z) \equiv e^{i\delta}$ *or* $f^{**}(z)$ *maps* G *onto a* ν*-sheeted disk*, $\nu \geqslant n$.

Using the theorems of §§3 and 4 and Theorem 6.3 of [55] and Theorems 1.1–1.3, we obtain the following.

THEOREM 1.6. *Let* Γ *consist of analytic contours and let the functions* $\omega_j(\zeta)$ *determining the functionals* (1.19) *be meromorphic in* $\overline{G}$ *(with poles in* G*). Let* $\beta_1, \dots, \beta_{m_1}$ *be the poles of* $\omega_1, \dots, \omega_m$*, counted with their multiplicities. Under the mapping* (1.19) *of the class* $E_q(G)$ $(1 \leqslant q \leqslant +\infty)$, *the image in* C^m *of the class* $E^1_{q,p}$ *is a convex closed bounded set* A_m *containing the origin in its interior. To each boundary point of* A_m *corresponds a function* $f^*(z)$ *of the same form as in* (6.13) *of* [55] *with* m *replaced by* m_1. *Uniqueness of* $f^*(z)$ *is guaranteed for* $q \neq 1$.

For any complex numbers $c_1, \dots, c_m$ *not all equal to zero, among all functions* $f(z) \in E_q$ *satisfying the interpolation conditions* (1.15) *the infimum*

$$\inf \int_\Gamma \left| \frac{f(\zeta)}{\rho(\zeta)} \right|^q ds \qquad (q < +\infty) \tag{1.24}$$

(*or* (1.22) *for* $q = +\infty$) *is attained at a function of the form* (6.13) *in* [55] (*with* m *replaced by* m_1). *This function is unique for* $q \neq 1$. *Concerning* $\rho(\zeta)$, *we assume that* $0 < m \leqslant \rho(\zeta) \leqslant M$ *for* $1 \leqslant q < +\infty$.

REMARK 1.7. In connection with the second part of Theorem 1.6, one should remember that the functions (6.13) in [55] already contain a constant factor c_1, and so there is no need to write down the solution of problem (1.24) in the form $f^*(z) = \lambda f^{**}(z)$.

Let us give one interesting application of the above theorems.

THEOREM 1.8. *Let* $f(z) \not\equiv \text{const}$ *be an arbitrary bounded analytic function in* G. *Then there exists a sequence* $\{\varphi_j(z)\}$ *converging to* $f(z)$ *uniformly on compact subsets of* G *and such that each function* $\varphi_j(z)$ *maps* G *onto a* ν_j*-sheeted disk of radius* $r = \sup|f(z)|$, *where* $n \leqslant \nu_j \leqslant n + j - 2$.

PROOF. Let z_0 be an arbitrary point in G. Let $f(z)$ have the following decomposition in a neighborhood of z_0:

$$f(z) = c_0 + c_1(z - z_0) + \cdots + c_j(z - z_0)^j + \cdots. \tag{1.25}$$

Consider the first j coefficients $c_0, \dots, c_{j-1}$. Representing Taylor coefficients by the Cauchy formula, we obtain linear functionals on $B(G)$ which have the form (1.19).

Among all functions in $B(G)$ which have the same first coefficients $c_0, \dots, c_{j-1}$ in a neighborhood of z_0, the smallest maximum modulus is attained at the function $f_j(z)$ mappsing G onto the ν_j-sheeted unit disk, where $n \leqslant \nu_j \leqslant n + j - 2$. The estimate of the number of sheets follows from Theorem 1.6; the total number of poles of the functions $\omega_k(\zeta)$, $k = 1, \dots, j-1$, is euqal to j. Let $r_j = \sup|f_j(z)|$. It is clear that $r_j \leqslant r$. Since all the functions $f_j(z)$ are uniformly bounded and their Taylor coefficients in the neighborhood of z_0 tend to the Taylor coefficients of $f(z)$ at z_0, then $f_j(z) \to f(z)$ uniformly inside G. It is also clear that $r_j \to r$. The sequence $\phi_j(z) = (r/r_j)f_j(z)$ satisfies all the requirements of our theorem.

§2. Carathéodory-Fejér problem

In Theorem 1.4–1.6 of the previous section we did not claim that *each* function of that form corresponds to boundary points of A_m or solves an extremal interpolation problem. In order to make such statements we have to specialize further the form of the linear functionals under consideration. Moreover, additional complications arise in case of a multiply connected domain, since functions of that form can be multivalued. In this section we state and prove more complete theorems for the unit disk and special l_j. We shall consider either functionals of the form

$$l_j(f) = f^{(j)}(\beta)/j!, \qquad j = 0, \dots, m-1, \tag{2.1}$$

where β is a fixed point inside the unit disk G, or

$$l_j(f) = f(\beta_j), \qquad j = 1, \dots, m, \tag{2.2}$$

where the β_j are fixed points in G. However, the results described below can be approximately extended to more general collections of functionals, for example, to

$$f(\beta_j),\ f'(\beta_j), \dots, f^{\nu_j}(\beta_j), \qquad j - 1, \dots, m,$$

where $\beta_1, \dots, \beta_m$ are given distinct points and the $\nu_j \geqslant 0$ are given integers. The body A_m will be constructed as in §1 using the functionals (2.1) or (2.2), and it will always be convex, closed, and bounded, with $(0, 0, \dots, 0)$ as an interior point. We will not repeat this in the statements of our theorems.

In case of the unit disk $H_q(G) = E_q(G)$ and in accordance with previous sections we furnish $H_q(G)$ with the norm

$$\|f\|_j = \left\{ \int_\Gamma \left| \frac{f(\zeta)}{\rho(\zeta)} \right|^q d\theta \right\}^{1/q}, \qquad \zeta = e^{i\theta}, \quad 1 \leqslant q < +\infty \tag{2.3}$$

(Γ is the unit circle) or

$$\|f\|_\infty = \operatorname*{ess\,sup}_{\zeta \in \Gamma} \left| \frac{f(\zeta)}{\rho(\zeta)} \right|. \tag{2.4}$$

The weight functions $\rho(\zeta)$ satisfy the same conditions as in the previous section.

THEOREM 2.1. *To every boundary point of the set A_m constructed with the aid of functionals* (1.1) *or* (2.2) *for the class $H^1_{\rho(\zeta),q}$ corresponds a function $f^*(z)$ of the form*

$$f^*(z) = C_1 \exp\left[\frac{1}{2\pi}\int_0^{2\pi} \ln \rho(e^{i\theta}) \frac{e^{i\theta}+z}{e^{i\theta}-z}\,d\theta\right] \prod_{j=1}^{k} \frac{z-\alpha_j}{1-\bar{\alpha}_j z}$$

$$\cdot \prod_{j=1}^{k} (1-\bar{\alpha}_j z)^{2/q} \cdot \prod_{j=1}^{k} (1-\beta_j z)^{-2/q} \qquad (k \leqslant m-1) \tag{2.5}$$

where $\alpha_1, \ldots, \alpha_{m-1}$ are points in the disk $|z| \leqslant 1$, and $\|f^*\|_q = 1$. In the case of functionals (2.1) *one must set $\beta_1 = \cdots = \beta_m = \beta$ in* (2.5). *Conversely, each function $f^*(z)$ of the form* (2.5) *corresponds to a boundary point of A_m. If $q \neq 1$, the function $f^*(z)$ corresponding to a given boundary point is unique.*

PROOF. To be specific, we will carry out the proof for the functionals (2.2). Let $f^*(z)$ have the form (2.5), $\|f^*\|_q = 1$. Construct the function

$$\omega(z) - \varphi^*(z) = C_2 \exp\left[-\frac{1}{2\pi}\int_0^{2\pi} \ln \rho(e^{i\theta}) \frac{e^{i\theta}+z}{e^{i\theta}-z}\,d\theta\right]$$

$$\cdot \prod_{j=1}^{m-1} (1-\bar{\alpha}_j z)^{2/p} \cdot \prod_{j=1}^{m-1} \frac{1-\bar{\beta}_j z}{z-\beta_j} \cdot \prod_{j=1}^{m-1} (1-\bar{\beta}_j z)^{-2/p} \tag{2.6}$$

where the points $\alpha_1, \ldots, \alpha_{m-1}$ are the same as in (2.5). It is easy to check that

$$\left|\int_\Gamma f^*(\zeta)[\omega(\zeta)-\varphi^*(\zeta)]\,d\zeta\right| = \int_\Gamma |f^*(\zeta)[\omega(\zeta)-\varphi^*(\zeta)]|\,|d\zeta|$$
$$= \left\{\int_\Gamma [\rho(\zeta)|\omega(\zeta)-\varphi^*(\zeta)|]^p\right\}^{1/p} \cdot \left\{\int_\Gamma \left|\frac{f^*(\zeta)}{\rho(\zeta)}\right|^q d\zeta\right\}^{1/q} \tag{2.7}$$

and, therefore, $f^*(z)$ is an extremal element of the functional

$$l(f) = \int_\Gamma f(\zeta)[\omega(\zeta)-\varphi^*(\zeta)]\,d\zeta. \tag{2.8}$$

From (2.6) it readily follows that

$$\omega(\zeta) - \varphi^*(\zeta) = \prod_{j=1}^{m} \frac{\lambda_j}{z-\beta_j} - \varphi_1(z), \tag{2.9}$$

where the λ_j are coefficients and $\varphi_1(z) \in E_p(G)$. Hence the functional (2.8) belongs to the linear span L of the functionals $l_1, \ldots, l_m$. Since $f^*(z)$ is an extremal element for the functional $l \in L$ and $\|f^*\| = 1$, according to Theorem 1.1, to the function $f^*(z)$ there corresponds a boundary point of A_m. The remaining part of Theorem 2.1 follows from Theorem 1.6.

THEOREM 2.2. *Let the numbers $c_1, \ldots, c_m$ be given. Among the functions $f(z) \in H_q(G)$, $q \geqslant 1$, such that*

$$l_j(f) = c_j, \qquad j = 1, \ldots, m, \tag{2.10}$$

there exists a unique function $f^*(z)$ *of the form* (2.5). *This is the only function among all the* $f(z)$ *satisfying* (2.10) *for which* $\min\|f\|_q$ *is attained. Moreover,*

$$\min\|f\|_q = \max_{\mathscr{F}\in H^1_{\rho_1,p}}\left|\sum_{j=1}^m c_j l_j(\mathscr{F})\right|\cdot 2\pi, \tag{2.11}$$

where

$$\begin{aligned}\rho_1(\zeta) &= \left(\Pi_1^m|\zeta-\beta_j|\right)/\rho(\zeta) \text{ for the functionals } (2.2),\\ \rho_1(\zeta) &= |\zeta-\beta|^m/\rho(\zeta) \text{ for the functionals } (2.1).\end{aligned} \tag{2.12}$$

To be specific, argue the case of functionals (2.2).

Construct a function $f_0(z)$ satisfying (2.10); for instance, for $f_0(z)$ we may take the Lagrange interpolation polynomial. If $f(z)$ is another function interpolating the values (2.10), then

$$f(z) = f_0(z) - \prod_{j=1}^m (z-\beta_j)\varphi(z) \qquad \forall\varphi\in H_q. \tag{2.13}$$

Therefore

$$\begin{aligned}\min\|f\|_q &= \min_{\varphi\in H_q}\left\{\int_\Gamma \rho_1(\zeta)^q\left|\sum_{j=1}^m \frac{c_j}{\zeta-\beta_j} - \varphi(\zeta)\right|^q |d\zeta|\right\}^{1/q}\\ &= \max_{\mathscr{F}\in H^1_{\rho_1,p}}\left|\int_\Gamma \mathscr{F}(\zeta)\left[\sum_{j=1}^m \frac{c_j}{\zeta-\beta_j}\right]d\zeta\right|\\ &= \max_{\mathscr{F}\in H^1_{\rho_1,p}}\left|\sum_{j=1}^m c_j l_j(\mathscr{F})\right| 2\pi.\end{aligned} \tag{2.14}$$

Here we have used the duality relations in §2 of [55].

As follows from the results in §§3 and 4 of [55], the function $\varphi(z)\in H_q$ giving the minimum in (2.14) is uniquely defined for every q (G is simply connected). In view of (2.13), this implies that the function $f^*(z)$ realizing $\min\|f\|_q$ is always unique. In particular, it is unique for $q = 1$. This was not guaranteed by Theorem 1.6, which deals with a more general situation. According to Theorem 1.6 this function must have the form (2.5). If there existed another function $f_1(z)$ of the form (2.5) interpolating the same conditions (2.10), then necessarily $\|f_1\| > \|f^*\|$ would be true. Consider the functions

$$f_2(z) = f_1(z)/\|f\|_q, \qquad f^{**}(z) = f^*(z)/\|f^*\|_q. \tag{2.15}$$

Since $f_2(z)$ and $f^{**}(z)$ have the form (2.5), boundary points of A_m correspond to them. But the point $c/\|f_1\| = (c_1/\|f_1\|,\dots,c_m/\|f_1\|)$ corresponds to the function $f_2(z)$ and the point $c/\|f^*\|$ to $f^{**}(z)$. These points are different, and we obtain an obvious contradiction to the structure of A_m. The proof is complete.

THEOREM 2.3. *Let* $c_0, c_1,\dots,c_{m-1}$ *be given numbers not all equal to zero. Let*

$$\lambda = \lambda(c_0,c_1,\dots,c_{m-1}) = \min\|f\|_\infty, \tag{2.16}$$

where the minimum is taken over all functions $f(z) \in H_\infty$ $(\rho \equiv 1)$ *satisfying the conditions*

$$f(0) = c_0,\ f'(0) = c_1, \ldots, f^{(m-1)}(0)/(m-1)! = c_{m-1}. \tag{2.17}$$

The number λ *is the largest positive root of the equation*

$$\mathscr{D}_m(\lambda) = \begin{vmatrix} \lambda & 0 & 0 & \cdots & 0 & c_0 & c_1 & \cdot & \cdot & c_{m-1} \\ 0 & \lambda & 0 & \cdots & 0 & 0 & c_0 & \cdot & \cdot & c_{m-2} \\ & & \ddots & & & & & c_0 & \cdots & c_{m-3} \\ & & & \ddots & & & & & \ddots & \cdot \\ & & & & \ddots & & & & & c_0 \\ \bar{c}_0 & 0 & \cdots & \cdots & 0 & \lambda & \ddots & & & \\ \bar{c}_1 & \bar{c}_0 & & & & & \ddots & & & \\ \cdot & \cdot & \ddots & & & & & \ddots & & \\ & & & \ddots & & & & & \ddots & \\ \bar{c}_{m-1} & \bar{c}_{m-2} & \cdot & \cdot & \bar{c}_0 & \cdot & \cdot & \cdot & \cdot & \lambda \end{vmatrix} = 0. \tag{2.18}$$

If

$$\lambda = \min \|f\|_\infty, \qquad f(\beta_j) = c_j, \quad j = 1, \ldots, m, \tag{2.19}$$

then λ *is the largest positive root of the equation*

$$\begin{vmatrix} \dfrac{c_1\bar{c}_1 - \lambda^2}{1 - \beta_1\bar{\beta}_1} & \dfrac{c_1\bar{c}_2 - \lambda^2}{1 - \beta_1\bar{\beta}_2} & \cdots & \dfrac{c_1\bar{c}_m - \lambda^2}{1 - \beta_1\bar{\beta}_m} \\ \dfrac{c_2\bar{c}_1 - \lambda^2}{1 - \beta_2\bar{\beta}_1} & \dfrac{c_2\bar{c}_2 - \lambda^2}{1 - \beta_2\bar{\beta}_2} & \cdots & \dfrac{c_2\bar{c}_m - \lambda^2}{1 - \beta_2\bar{\beta}_m} \\ \cdots & \cdots & \cdots & \cdots \\ \dfrac{c_m\bar{c}_1 - \lambda^2}{1 - \beta_m\bar{\beta}_1} & \dfrac{c_m\bar{c}_2 - \lambda^2}{1 - \beta_m\bar{\beta}_2} & \cdots & \dfrac{c_m\bar{c}_m - \lambda^2}{1 - \bar{\beta}_m\beta_m} \end{vmatrix} = 0. \tag{2.20}$$

PROOF. Consider problem (2.16), subject to conditions (2.17). According to Theorem 2.2, the unique extremal function has the form (2.5). For $q = \infty$ or 1, (2.5) transforms into

$$f^*(z) = C\prod_{j=1}^{k} \frac{z - \alpha_j}{1 - \bar{\alpha}_j z}, \qquad k \leqslant m - 1. \tag{2.21}$$

The function (2.21) can be rewritten in the form

$$f^*(z) = \lambda\frac{\bar{p}_k + \bar{p}_{k-1}z + \cdots + \bar{p}_0 z^k}{p_0 + p_1 z + \cdots + p_k z^k}, \qquad \lambda > 0. \tag{2.22}$$

Since $\lambda = |f^*(e^{i\theta})| = \|f\|_\infty$, we have $\lambda = \lambda(c_0, \ldots, c_{m-1})$. If $k < m - 1$, then, multiplying the numerator and denominator by $1 + z + \cdots + z^{m+k}$, we reduce (2.22) to the form

$$f^*(z) = \frac{\bar{p}_{m-1} + \bar{p}_{m-2}z + \cdots + \bar{p}_0 z^{m-1}}{p_0 + p_1 z + \cdots + p_{m-1}z^{m-1}} = c_0 + c_1 z + \cdots + c_{m-1}z^{m-1} + \cdots. \tag{2.23}$$

Comparing coefficients of like powers of z, we obtain the following system of equations:

$$\begin{aligned} \lambda \bar{p}_{m-1} &= c_0 p_0, \\ \lambda \bar{p}_{m-2} &= c_0 p_1 + c_1 p_0, \\ \cdots &\cdots \\ \lambda \bar{p}_0 &= c_0 p_{m-1} + \cdots + c_{m-1} p_0. \end{aligned} \tag{2.24}$$

Let us add to this system the equations obtained from (2.24) by replacing all terms by their conjugates. We obtain a homogeneous system of $2m$ linear equations with the unknowns $\bar{p}_{m-1}, \ldots, \bar{p}_0, p_0, \ldots, p_{m-1}$. Since this system has a nonzero solution, namely the coefficients of the fraction (2.23), the determinant $\mathcal{D}_m(\lambda)$ of the system vanishes. $\mathcal{D}_m(\lambda)$ is the characteristic determinant of a hermitian matrix and has only real roots. Some of them are positive, e.g., the number $\lambda = \lambda(c_0, \ldots, c_{m-1})$. (Note that the roots of $\mathcal{D}_m(\lambda)$ are pairwise symmetric.) Let us show that $\lambda(c_0, \ldots, c_{m-1})$ is the largest positive root. Let λ' be a different positive root of (2.18), $\lambda' > \lambda(c_0, \ldots, c_{m-1})$. From system (2.24) and its conjugate with $\lambda > \lambda'$ we find the unknowns $\bar{p}_{m-1}, \ldots, \bar{p}_0, p_0, \ldots, p_{m-1}$. But we need those values of the unknowns for which p_k and $\bar{p}_k$ are mutually conjugate. If $\bar{s}_{-(m-1)}, \bar{s}_{-(m-2)}, \ldots, \bar{s}_{-0}, s_0, \ldots, s_{m-1}$ are any nonzero values of unknowns in the system, then $\bar{s}_{m-1}, \bar{s}_{m-2}, \ldots, \bar{s}_0, s_{-0}, \ldots, s_{-m-1}$ also satisfy the same system. Setting $p_k = s_k + \bar{s}_k$, $k = -(m-1), \ldots, -0, 0, \ldots, m-1$, or $p'_k = i(s_k - s_{-k})$, $k = -(m-1), \ldots, -0, 0, \ldots, m-1$, we obtain solutions which satisfy the required conjugacy condition. Also, one of the solutiosn $\{p_k\}$ or $\{p'_k\}$ is certainly nontrivial. Otherwise, the solution $\{s_k\}$ would have been trivial. Assume that $\{p_k\}$ is a nontrivial solution.

Using the numbers p_k be construct a fraction $R(z)$ of the form (2.23) with $\lambda = \lambda'$. Since $|R(e^{i\theta})| = \lambda'$, $R(z)$ has neither zeros nor poles on the unit circle Γ. At the same time, according to the uniqueness of the fraction (2.23) analytic in the unit disk and interpolating (2.17), the fraction $R(z) \not\equiv f^*(z)$ must have poles inside the disk. Consider the difference $\phi(z) = R(z) - f^*(z)$. The number τ of zeros of that difference in the disk $|z| < 1$ satisfies the inequality $\tau \geqslant m$. The number σ of poles is the same for $\varphi(z)$ and $R(z)$. On the unit circle we have

$$|\varphi(e^{i\theta})| \geqslant |R(e^{i\theta})| - |f^*(e^{i\theta})| = \lambda' - \lambda > 0.$$

Therefore, $\Delta_\Gamma \arg \varphi(e^{i\theta}) = 2\pi(\tau - \sigma)$. On the other hand, since $|R(e^{i\theta})| > |f^*(e^{i\theta})|$, $\Delta_\Gamma \arg \varphi = \Delta_\Gamma \arg R = 2\pi(t - \sigma)$, where t is the number of zeros of $R(z)$ in the unit disk. Hence $t \leqslant m - 1 < \tau$. We have arrived at a contradiction. Therefore, there is no $\lambda' > \lambda(c_0, \ldots, c_{m-1})$.

We have proved the first part of the theorem. The second assertion can be established by a similar argument. However, we should seek the required fraction in the form

$$f(z) = \sum_{k=1}^{m} \frac{A_k c_k}{z - \beta_k} \Big/ \sum_{k=1}^{m} \frac{A_k}{z - \beta_k}.$$

Since the modulus of $f(z)$ on $|z| = 1$ must be constant (equal to λ), we have, according to the symmetry principle, $f(1/\bar{\beta}) = \lambda^2/\bar{c}_j$. From this we obtain the following systems of equations to find the A_k:

$$\sum_{k=1}^{m} A_k \frac{c_k\bar{c}_j - \lambda^2}{\beta_k\bar{\beta}_j - 1} = 0.$$

§3. On Taylor coefficients of H_∞- and H_1-functions

In this section we consider some questions concerning coefficients of functions in H_p, $p \geqslant 1$, in the unit disk. We will mainly concentrate on the class H_p and then on H_1. Let

$$l_j(f) = \frac{f^{(j)}(0)}{j!}, \qquad j = 0, 1, \ldots, m - 1. \tag{3.1}$$

For the class $H_\infty^1 = B^1$ we consider the mapping into the space $\mathbf{C}^m$ defined by the functionals (3.1). As usual we denote the resulting set by A_m. The following theorem answers the question of what the numbers $c_0, \ldots, c_{m-1}$ should be for the polynomial $c_0 + c_1 z + \cdots + c_{m-1}z^{m-1}$ to be a Taylor polynomial of a function $f \in H_\infty^1$ at zero. Let $\mathscr{D}_k(\lambda)$, $k = 1, 2, \ldots,$ denote the determinant of the form (2.18).

THEOREM 3.1. *For $(c_0, c_1, \ldots, c_m)$ to be an interior point of A_m, it is necessary and sufficient that*

$$\mathscr{D}_k(1) > 0, \qquad k = 1, \ldots, m - 1, m. \tag{3.2}$$

For $(c_0, c_1, \ldots, c_{m-1})$ to be a boundary point of A_m it is necessary and sufficient that for some $k_0, 1 \leqslant k_0 \leqslant m$,

$$\begin{gathered} \mathscr{D}_k(1) > 0 \quad \textit{for } k = 1, \ldots, k_0 - 1, \\ \mathscr{D}_{k_0}(1) = \mathscr{D}_{k_0+1}(1) = \cdots = \mathscr{D}_m(1) = 0, \end{gathered} \tag{3.3}$$

and also that all principal minors of $\mathscr{D}_m(1)$ be nonnegative.

PROOF. Let $(c_0, c_1, \ldots, c_{m-1})$ be an interior point of A_m. Then $\lambda^* = \lambda(c_0, \ldots, c_{m-1}) < 1$. Let H be a hermitian quadratic form of the variables $x_1, \ldots, x_{2m}$ whose matrix coincides with the matrix of $\mathscr{D}_m(\lambda)$ from (2.18). Let h denote the positive hermitian form $h = \sum_1^{2m} x_j\bar{x}_j$. Consider a pencil of quadratic forms $H - \lambda h$ (see [9], Chapter X, §§8–9). $-\lambda^*$ is the smallest characteristic number of that pencil, i.e., the smallest root of the determinant $|H - \lambda h| = D_m(\lambda) = \mathscr{D}_m(-\lambda)$. Since the smallest characteristic number is the minimum of the ratio H/h, the form $H + \lambda^* h$ is nonnegative. Hence, $H + h$ is positive. Let us apply Sylvester's criterion for positivity of a hermitian form ([9], Chapter X, §9, Theorem 19). Writing down the determinants which appear in that criterion, we verify that m of them are simply equal to one (they all have ones on the main diagonal and zeros elsewhere). The remaining m determinants coincide with $\mathscr{D}_k(1)$, $k = 1, \dots, m$, after obvious transformations. Thus, according to Sylvester's criterion, (3.3) holds.

Now, if (3.3) holds, then the form $H + h$ is positive. Therefore $\min H/h = -\lambda^* > -1$. Hence, $\lambda(c_0, \dots, c_{m-1}) = \lambda^* < 1$, and $(c_0, \dots, c_{m-1})$ is an interior point of A_m.

If $(c_0, \dots, c_{m-1}) \in \partial A_m$, then $\lambda^* = \lambda(c_0, \dots, c_{m-1}) = 1$. Arguing as we did above, we conclude that the form $H + h$ is nonnegative, and so (see [9], Chapter X, §9, Theorem 20, and [10]) all principal minors of $\mathscr{D}_m(1)$ are nonnegative. In particular, $\mathscr{D}_k(1) \geqslant 0$, $k = 1, \dots, m$. The strict inequalities (3.2) cannot hold, since this would imply that $(c_0, \dots, c_{m-1})$ lies in the interior of A_m. Therefore, $\mathscr{D}_{k_0}(1) = 0$ for some $k_0, 1 \leqslant k_0 \leqslant m$. This means that $1 = \lambda(c_0, \dots, c_{m-1})$. Since $\lambda(c_0, \dots, c_{k_0-1}) \leqslant \lambda(c_0, \dots, c_{k_0}) \leqslant \lambda(c_0, \dots, c_{m-1}) = 1$, it follows that

$$\lambda(c_0, \dots, c_{k_0-1}) = \lambda(c_0, \dots, c_{k_0}) = \lambda(c_0 = c_{m-1}) = 1.$$

Hence, (3.3) holds. Conversely, if (3.3) holds and all principal minors of $\mathscr{D}_m(1)$ are nonnegative, then $H + h \geqslant 0$. Therefore, $-\lambda^* \geqslant -1$ and $\lambda^* \leqslant 1$. Hence, $(c_0, c_1, \dots, c_{m-1}) \in A_m$, and since (3.2) is not satisfied, $(c_0, \dots, c_{m-1}) \in \partial A_m$.

It is easy to verify that (3.3) alone does not guarantee that the points $(c_0, \dots, c_{m-1})$ belong to A_m. Set, for example, $c_0 = 1$. Clearly there exists only one function $f(z) \equiv 1$ in the class B^1 such that $f(0) = 1$.

Considering the equation $\mathscr{D}_2(1) = 0$ with $c_0 = 1$ and unknown c_1, we obtain the unique solution $c_1 = 0$. If we now consider the equation $\mathscr{D}_3(1) = 0$ with $c_0 = 1$, $c_1 = 0$ and unknown c_2, then it is satisfied for any c_2, and not only for $c_2 = 0$. Thus, $\mathscr{D}_1(1) > 0$, $\mathscr{D}_2(1) = 0$ and $\mathscr{D}_3(1) = 0$ for any point $(1, 0, c_2)$. Nevertheless, the following theorem is true.

THEOREM 3.2. *For a sequence*

$$c_0, c_1, \dots, c_{m-1}, \dots \tag{3.4}$$

to be a sequence of the Taylor coefficients of a function $f(z) \in B^1$ it is sufficient and necessary that

$$\mathscr{D}_j(1) > 0, \qquad j = 1, 2, \dots, \tag{3.5}$$

or

$$\mathscr{D}_j(1) > 0, \quad j = 1, \dots, m, \qquad \mathscr{D}_j(1) = 0, \quad j = m+1, \dots. \tag{3.6}$$

In the latter case (and only in that case) $f(z)$ has the form of the function (1.21′), $k \leqslant m - 1$, *and this fraction is irreducible.*

The proof involves overcoming certain algebraic difficulties, and we omit it.

An interesting addition to Theorem 3.1 is given by the following theorem which describes the behavior of the coefficient c_m of a function $f(z) \in B^1$ when the values of the coefficients $c_0, \dots, c_{m-1}$ are given.

Given $c_0, \ldots, c_{m-1}$, construct the determinants $\mathscr{D}_k(1)$, $k = 1, \ldots, m$, and also a determinant $\mathscr{D}_{m+1}(1)$ with the same $c_0, \ldots, c_{m-1}$ and a variable c_m. To emphasize the dependence on c_m, we denote this determinant by $\mathscr{D}_{m+1}(1, c_m)$. Let $\mathscr{D}^*_{m+1}(1, c_m)$ denote the determinant obtained by deleting the last column and the first row from $\mathscr{D}_{m+1}(1, c_m)$. Also, let $\mathscr{D}^{**}(1, c_m)$ stand for the determinant obtained by deleting the first column and the last row from $\mathscr{D}_{m+1}(1, c_m)$. It is clear that $\mathscr{D}^*_{m+1}(1, c_m) = \overline{\mathscr{D}^{**}_{m+1}(1, c_m)}$.

THEOREM 3.3. *Let $(c_0, c_1, \ldots, c_{m-1})$ be such that* (3.2) *holds. Let C_m be the domain of the coefficient c_m obtained from all possible choices of function $f(z)$ whose Taylor expansion starts with the polynomial $c_0 + c_1 z + \cdots + c_{m-1} z^{m-1}$. Then C_m is the disk with center*

$$\zeta_0 = \frac{\mathscr{D}^*_{m+1}(1, 0)}{\mathscr{D}_{m+1}(1)} \tag{3.7}$$

and radius

$$z = \frac{\mathscr{D}_m(1)}{\mathscr{D}_{m-2}(1)}. \tag{3.8}$$

PROOF. For $(c_0, c_1, \ldots, c_{m-1}, c_m)$ to be an interior point of A_m, it is necessary and sufficient (in view of (3.2) and Theorem 3.1) that $\mathscr{D}_{m-1}(1, c_m) > 0$. Using a well-known theorem on minors of a determinant (see, for instance, [9], Chapter II, §3), we can write

$$\mathscr{D}_{m+1}(1, c_m)\mathscr{D}_{m-1}(1) = \mathscr{D}^2_m(1) - \mathscr{D}^*_{m+1}(1, c_m)\mathscr{D}^{**}(1, c_m).$$

Since $\mathscr{D}_{m-1}(1) > 0$, the inequality we need is equivalent to

$$\mathscr{D}^2_m(1) - \mathscr{D}^{**}_{m+1}(1, c_m)\,\overline{\mathscr{D}^{**}_{m+1}(1, c_m)} > 0. \tag{3.9}$$

Now, taking into account the way $\mathscr{D}^{**}_{m+1}(1, c_m)$ depends on c_m, it is easy to see that (3.9) is equivalent to the inequality $|c_m - \zeta_0| < r$, where ζ_0 and r are defined by (3.7) and (3.8). It is clear that if the point c_m lies on the boundary of that disk, then $(c_0, \ldots, c_{m-1}, c_m)$ lies on the boundary of A_m.

Let us give another method of investigating the question of whether (3.4) is a sequence of coefficients of a function in B^1. Set

$$f(z) = \sum_{n=0}^{\infty} c_n z^n \tag{3.10}$$

and define the functions

$$f_k(z) = \frac{1}{2}\,\frac{f_{k-1}(z) - f_{k-1}(0)}{1 - \overline{f_{k-1}(0)}\,f_{k-1}(z)}, \qquad k = 1, 2, \ldots,$$
$$f_0(z) = f(z), \tag{3.11}$$

and the numbers

$$a_k = f_k(0), \qquad k = 0, 1, \ldots. \tag{3.12}$$

THEOREM 3.4. *If $f(z) \in B^1$, then either*

$$|a_k| < 1, \qquad k = 0, 1, \ldots, \tag{3.13}$$

or, for some $n \geqslant 0$,

$$|a_k| < 1, \quad k = 0, 1, \ldots, n-1, \qquad |a_n| = 1. \tag{3.14}$$

The latter takes place only if $f(z)$ is a finite Blaschke product (2.2) *with $k = n$. Conversely, if* (3.13) *or* (3.14) *holds, then $f(z) \in B^1$.*

PROOF. If $\varphi(z) = \sum_1^\infty b_k z^k \in B^1$, then from the Cauchy formulas it follows that $|b_k| \leqslant 1$, $k = 0, 1, \ldots$. If $|b_0| = 1$, then $\varphi(z) \equiv b_0$. If in our case $|a_0| = 1$, then $f_0(z) = a_0$ and (3.14) holds together with (2.21), where $k = 0$. If $|a_0| < 1$, then $f_1(z) \in B^1$ and, hence, $|a_1| \leqslant 1$. If $|a_1| = 1$, then $f_1(z) \equiv a_1$, and expressing $f(z)$ in terms of $f_1(z)$ we obtain (3.14) and (1.21′) with $k = 1$. If $|a_1| < 1$, then $f_2(z) \in B^1$, and so on.

Let $|a_k| < 1$, $k = 0, \ldots, n-1$. So $f_0, \ldots, f_{n-1}$ belongs to B^1. Then $f_n(z) \in B^1$ and, therefore, $|a_n| \leqslant 1$. There are only two possibilities: either (3.13) holds, or, for some n, $|a_n| = 1$ for the first time. Then $f_n(z) \equiv a_n$ and, expressing f_{n-1} in terms of a_n, f_{n-2} in terns of f_{n-1}, etc., in accordance with (3.11), we obtain the formula (2.21) for $f(z)$ with $k = n$.

We have proved the theorem in one direction. For subsequent use it is worthwhile to write down the inversions of formulas (3.11):

$$f_{k-1}(z) = \frac{a_{k-1} + z f_k(z)}{1 + \bar{a}_{k-1} z f_k(z)}. \tag{3.15}$$

Let us prove the sufficiency of our conditions. Suppose (3.14) holds. Again let us express f_{n-1} in terms of a_n, noting that $f_{n-1} \in B^1$. Then, we express f_{n-2} in terms of f_{n-1} and again $f_{n-2}(z) \in B^1$, and so on until we arrive at $f(z)$ and $f(z) \in B^1$. If (3.13) holds, then for a fixed n let us consider the numbers $a_0, a_1, \ldots, a_{n-1}, 1$. As above, from these numbers we construct a fraction $\varphi(z) \in B^1$ which has the form (2.21). The first n Taylor coefficients of $f(z)$ and $\varphi(z)$ coincide. Indeed, for the functions $\varphi_k(z)$ we have

$$\varphi_n(z) \equiv 1, \qquad \varphi_{k-1}(z) = \frac{a_{k-1} + z\varphi_k(z)}{1 + \bar{a}_{k-1} z \varphi_k(z)}, \quad k = 1, \ldots, n-1. \tag{3.16}$$

Comparing (3.15) and (3.16), we find that

$$f_{k-1}(z) - \varphi_{k-1}(z) = \frac{\left(1 - |a_{k-1}|^2\right) z \left(f_k(z) - \varphi_k(z)\right)}{\left(1 + \bar{a}_{k-1} z f_k(z)\right)\left(1 + \bar{a}_{k-1} z \varphi(z)\right)}. \tag{3.17}$$

Therefore,

$$f(z) - \varphi(z) = z^{n-1} Q(z) \left[f_{n-1}(z) - \varphi_{n-1}(z) \right], \tag{3.18}$$

where $Q(z)$ is analytic in the disk and $f_{n-1}(0) - \varphi_{n-1}(0) = a_{n-1} - a_{n-1} = 0$.

To finish the proof it remains to apply the compactness principle.

Theorems 3.1 and3.2 contain, in particular, the criterion for $f(z) \in B^1$ to be a finite Blaschke product. If we consider now not all the bounded functions, but only inner functions, then using the duality of extremal problems we can give another sufficient criterion for a function to be a finite Blaschke product. This criterion also allows us to obtain estimates of the coefficients.

THEOREM 3.5. *Let $f(z) = \sum_1^\infty c_n z^n$ be an inner function. Denote by A_n the infinite matrix*

$$A_n = \begin{pmatrix} |c_n| & |c_{n+1}| & |c_{n+2}| & \cdots \\ |c_{n+1}| & |c_{n+2}| & |c_{n+3}| & \cdots \\ |c_{n+2}| & |c_{n+3}| & |c_{n+4}| & \cdots \\ \cdots & \cdots & \cdots & \cdots \end{pmatrix}. \tag{3.19}$$

If for some n the quadratic form with the matrix A_n is bounded by a number less than 1, then $f(z)$ is a finite Blaschke product.

PROOF. Suppose that $f(z)$ is not a finite Blaschke product. Then there exists a sequence $\{z_n\}$ with $|z_n| \to 1$ and $f(z_n) \to 0$. This is clear if $f(z)$ contains an infinite Blaschke product as a factor. If not, then $f(z) = B(z)Q(z)$, where $B(z)$ is a finite Blaschke product, and

$$Q(z) = \exp \int_0^{2\pi} \frac{e^{i\theta} + z}{e^{i\theta} - z}\, d\mu,$$

with a singular measure $d\mu < 0$. There always exist points (see [48], Chapter IV, §9) where $\mu'(\theta) = -\infty$. Then $Q(z) \to 0$ as z tends to the boundary along radii pointing to those points. From this property of $f(z)$ it follows that

$$\inf_{g \in H_\infty} \| z^n - f(z) g(z) \|_\infty = 1. \tag{3.20}$$

Writing the dual of (3.20), we verify that for any $\varepsilon > 0$ there exists a function $h(z) = h_\varepsilon(z) \in H_1$ such that

$$\|h\|_1 = \int_0^{2\pi} |h(e^{i\theta})|\, d\theta$$

and

$$\left| \frac{1}{2\pi} \int_0^{2\pi} \zeta^n \overline{f(\zeta)}\, d\zeta \right| > 1 - \varepsilon, \qquad \zeta = e^{i\theta}. \tag{3.21}$$

Any function $h(z) \in H_1$ can be represented in the form $h(z) = h_1(z) \cdot h_2(z)$, where $h_i(z) \in H_2$ and $\|h_i\|_2 = \|h\|_1$, $i = 1, 2$. (This immediately follows from the Smirnov parametric representation.) Setting

$$h(z) = \sum_{n=0}^\infty b_n z^n, \qquad h_1(z) = \sum_{n=0}^\infty p_n z^n, \qquad h_2(z) = \sum_{n=0}^\infty q_n z^n, \tag{3.22}$$

we have

$$b_n = \sum_{k=1}^n p_k q_{n-k}, \qquad \sum_{n=0}^\infty |p_n|^2 = \sum_{n=0}^\infty |q_n|^2 = 1. \tag{3.23}$$

The left-hand side in (3.21) is equal to

$$|b_0\bar{c}_{n+1} + b_1\bar{c}_{n+2} + \cdots| = \left| \sum_{k=0}^{\infty} \bar{c}_{n+k+1} \sum_{i+j=k} p_i q_j \right|. \tag{3.24}$$

The last expression is a bilinear form in p_i and q_j, and the absolute values of its coefficients coincide with the elements of the matrix A_{n+1}. It is clear that the upper bound of the form (3.24) canno exceed the upper bound of the form whose matrix is A_{n+1}. But the upper bound of (3.24) is not smaller than one, and we arrive at the desired contradiction. The theorem is proved.

THEOREM 3.6. *If $f(z) = \Sigma_0^\infty c_n z^n$ is an inner function and*

$$\overline{\lim}\, n|c_n| < 1/\pi, \tag{3.25}$$

then $f(z)$ is a finite Blaschke product.

We omit the proof.

Going over to the H_p-classes, let us first recall the following corollary of the well-known Riesz-Fischer theorem (see, for instance, [4]).

THEOREM 3.7. *For $f(z) = \Sigma_0^\infty c_n z^n$ to belong to H_2, it is necessary and sufficient that*

$$\sum_{n=0}^{\infty} |c_n|^2 < \infty. \tag{3.26}$$

We have already used this simple fact in the proof of Theorem 3.5.

It is worthwhile to mention the following simple characterization of inner functions.

THEOREM 3.8. *For $f(z) = \Sigma_0^\infty c_n z^n$ to be an inner function it is necessary and sufficient that*

$$\sum_{n=0}^{\infty} |c_n|^2 = 1, \tag{3.27}$$

$$\sum_{n=0}^{\infty} c_n \bar{c}_{k+n} = 0, \qquad k = 1,2,\ldots. \tag{3.28}$$

PROOF. An inner function $f(z)$ belongs to H_2, and (3.27) holds. Moreover, as $|f(\zeta)| = 1$ almost everywhere,

$$0 = \frac{1}{2\pi}\int_0^{2\pi} \zeta^k\, d\theta = \frac{1}{2\pi}\int_0^{2\pi} \zeta^k |f(\zeta)|^2\, d\zeta, \qquad k = 1,2,\ldots, \tag{3.29}$$

and (3.29) immediately leads to (3.28). Conversely, if (3.27) holds, then $f(z) \in H_2$ and (3.29) imply that $|f(\zeta)| \equiv \text{const} \equiv 1$.

Let us describe the set of coefficients of functions in H_1. We construct the set A_m associated with coefficients $(c_0, \ldots, c_{m-1})$ of H_1-functions as has been done for H_∞^1.

THEOREM 3.9. *In order that* $(c_0, c_1, \ldots, c_m) \in A_m$ *it is necessary and sufficient that there exist numbers* $p_0, p_1, \ldots, p_{m-1}$ *such that*

$$2\pi \sum_{j=0}^{m-1} |p_j|^2 \leqslant 1, \qquad c_k = \sum_{j=0}^{k} p_j p_{k-j}, \quad k = 0, \ldots, m-1. \tag{3.30}$$

Moreover, if $(c_0, \ldots, c_{m-1}) \in \partial A_{m-1}$, *then the first inequality in* (3.30) *becomes an equality.*

PROOF. *Necessity.* If $(c_0, \ldots, c_{m-1}) \in \partial A_m$, then according to Theorems 1.1 and 2.1 and Corollary 6.7 in [55], to this point corresponds the polynomial $f(z) = (p_0 + p_1 z + \cdots + p_{m-1} z^{m-1})^2$. Also, $\|f\|_1 = 1$, and this immediately yields (3.30) with equality. If $(c_0, \ldots, c_{m-1})$ is an interior point of A_m, then it is a boundary point of the set λA_m for a certain λ, $0 \leqslant \lambda \leqslant 1$. The same argument as above leads to (3.30).

Sufficiency. If (3.30) holds, then the function $(p_0 + p_1 z + \cdots + p_{m-1} z^{m-1})^2$ belongs to H_1 and has the prescribed coefficients c_k, $k = 0, \ldots, m-1$.

§4. The problem of the radius of p-valence

Let G be an n-connected domain bounded by nondegenerate contours. (Without loss of generality we can assume that they are analytic.) Fix a point $z_0 \in G$ and complex numbers A_0 and A_1. Denote by $B(A_0 A_1)$ the subclass of $B^1(G)$ consisting of those $f(z)$, $|f(z)| < 1$, for which

$$f(z_0) = A_0, \qquad f'(z_0) = A_1. \tag{4.1}$$

We assume, of course, that A_0 and A_1 are chosen so that the class $B(A_0, A_1)$ is nonempty. A_0 can be arbitrary as long as $|A_0| < 1$. For A_1, Corollary 11.2 in [55] gives the estimate

$$|A_1| \leqslant \left(1 - |A_0|^2\right) F'(z_0, z_0). \tag{4.2}$$

We recall that $F(z, z_0)$ is the Ahlfors function of the point z_0. If we have equality in (4.2), then from the same corollary it follows that $B(A_0, A_1)$ consists of a single function. So, we assume strict inequality in (4.2).

THEOREM 4.1. *For any integer* $p \geqslant 1$ *all functions of the family* $B(A_0, A_1)$, *where*

$$|A_0| < 1, \qquad 0 < |A_1| < \left(1 - |A_0|^2\right) F'(z_0, z_0), \tag{4.3}$$

are at most p-valent in some disk $|z - z_0| < r_p$, *whose closure lies in the interior of the domain G. Moreover, there exist functions in this family (we call them extremal) such that each one of them is at least* $(p+1)$*-valent in every larger concentric circle which belongs to G. Each extremal function gives a conformal mapping of G onto the k-sheeted unit disk and*

$$\max(n, p+1) \leqslant k \leqslant n + p + 1. \tag{4.4}$$

PROOF. First let us show that $B(A_0, A_1)$ contains a function which is at least $(p+1)$-valent in some closed disk with center z_0 lying inside of G. To do this let us join z_0 to the closest point of the boundary Γ of G by a straight-line segment.

Take an arbitrary point $\tau \neq z_0$, $\tau \in G$, on this segment. Consider functions

$$f_1(z) = \varepsilon\left(\frac{F(z, z_0) - F(\tau, z_0)}{1 - \overline{F(\tau, z_0)}\, F(z, z_0)}\right)^p F(z, z_0), \qquad \varepsilon = e^{i\theta}. \tag{4.4'}$$

$$f_2(z) = \frac{f_1(z) + A_0}{1 + \overline{A_0} f_1(z)}. \tag{4.4''}$$

The function $f_2(z)$ belongs to $B^1(G)$, and $f_2(z_0) = A_0$. This function takes the value A_0 in the disk $|z - z_0| < |\tau - z_0|$ at least $p + 1$ times. Also,

$$f'(z_0) = \varepsilon\left(1 - |A_0|^2\right) \cdot M \cdot (-F(\tau, z_0))^p,$$

where $M = F'(z_0, z_0)$. As τ varies along the segment, $|F(\tau, z_0)|$ is a continuous function of τ with $|F(z_0, z_0)| = 0$ and $|F(\tau, z_0)| \to 1$ as $\tau \to \Gamma$. Since (4.3) holds, for some $\tau = \tau_0 \in G$ we have $|f_2'(z_0)| = |A_1|$. Choosing an appropriate ε, we obtain $f_2'(z_0) = A_1$. Thus, we have constructed a function $f_2(z) \in B(A_0, A_1)$ which takes the value A_0 in the disk $|z - z_0| < |\tau - z_0|$ (in the interior of G) at least $p + 1$ times, as required.

From the condition $f_2'(z) = A_1 \neq 0$, compactness of the family $B(A_0, A_1)$ and existence in that family of the above function $f_2(z)$, it follows that there exist a disk described in the theorem and functions called extremal.

Let us go over to the proof of the main part of the theorem, i.e., establishing the claimed geometric property of extremal functions.

Let $f_0(z)$ be an extremal function. This means that in the closed disk $|z - z_0| \leqslant r_p$, $f_0(z)$ takes a certain value at least $p + 1$ times (counting multiplicities of course). Therefore, there exist points $z_1, \ldots, z_s$, $1 \leqslant s \leqslant p + 1$, in this disk, different from each other and such that $f_0(z_1) = \cdots = f_0(z_s)$. Moreover, if $s < p + 1$, there exist points $z_1, \ldots, z_\nu$, $1 \leqslant \nu \leqslant s$, among them such that

$$f_0'(z_1) = \cdots = f_0^{(n_1)}(z_1) = 0, \quad \ldots, \quad f_0'(z_\nu) = \cdots = f_0^{n_\nu}(z_\nu) = 0,$$

where $n_1 \geqslant 1, \ldots, n_\nu \geqslant 1$ and $n_1 + \cdots + n_\nu + s = p + 1$. We consider $s < p + 1$ as the more general case. Define a map $P = T(f)$ of the class $B^1(G)$ into $(p + 2)$-dimensional complex space, consisting of points of the form $P = (c_0, c_1, d_1, \ldots, d_{s-1}, d_{11}, d_{12}, \ldots, d_{n_\nu \nu})$, as follows:

$$c_0 = f(z_0), \quad c_1 = f'(z_0), \qquad d_j = f(z_{j+1}) - f(z_1), \quad j = 1, \ldots, s - 1,$$

$$d_{jl} = f^{(l)}(z_j), \quad j = 1, \ldots, \nu, \quad l = 1, \ldots, n_\nu. \tag{4.5}$$

According to Theorem 1.6, the image of $B^1(G)$ under the map $T(f)$ is a convex closed solid A containing the origin as an interior point. To any boundary point of this set there corresponds a unique function which, if not a constant, maps G onto a k-sheeted disk, where

$$n \leqslant k \leqslant n + p + 1. \tag{4.6}$$

Let us show that the point $P_0 = T(f_0) = (A_0, A_1, 0, \ldots, 0)$ lies on the boundary of A. This would complete the proof of the theorem. Indeed, the case $f_0(z) \equiv \text{const}$ cannot occur; and, also, $f_0(z)$ is at least $(p + 1)$-valent in G. Hence, we can replace the estimate (4.6) for $f_0(z)$ by (4.4).

Assume that P_0 is an interior point of A. Denote by $\mathcal{U}$ a ball with center at P_0 which lies in the interior of A. Let $t > 0$ be the radius of $\mathcal{U}$. Take points $z_1', \dots, z_s'$ inside the disk $|z - z_0| < r_p$. We define a map $P' = T'(f) = (c_0, c_1, d_1', \dots, d_{\nu n_\nu}')$ of the class $B^1(G)$ into $(p+2)$-dimensional space by (4.5), with the points $z_0, z_1, \dots, z_s$ replaced by $z_0, z_1', \dots, z_s'$ respectively. Take $z_1', \dots, z_s'$ so close to $z_1, \dots, z_s$ that $\rho(T(f), T'(f)) < t/2$ for all $f \in B^1(G)$. Here $\rho(P, P')$ is the distance between P and P'.

The image of $B^1(G)$ under the mapping $T'(f)$ is a convex body A'. Let us show that the point $P_0 = T(f_0) \in A'$. If $P_0 \notin A'$, then we can find a hyperplane π passing through P_0 such that the whole of A' lies in the same half-space defined by π. Drop a perpendicular from P_0 to π pointing away from the half-space containing A'. Let P_1 denote the intersection point of that perpendicular with the boundary of the ball $\mathcal{U}$. It is clear that the distance from P_1 to any point of A' is larger than t. But if $P_1 = T(f_1)$, then according to our construction $\rho(T(f_1), T'(f_1)) < t/2$. This contradiction proves that $P_0 \in A'$.

Hence, there exists a function $f_2(z) \in B^1(G)$ such that $P_0 = T'(f_2)$. This yields the following equalities:

$$f_2(z_0) = A_0, \quad f_2'(z_0) = A_1, \quad f_2(z_{j+1}') - f(z_1) = 0, \quad j = 1, \dots, s-1,$$
$$f_2'(z_1') = \cdots = f_2^{(n_1)}(z_1') = 0, \quad \dots, \quad f_2'(z_\nu^1) = \cdots = f_2^{(n_\nu)}(z_1') = 0. \quad (4.7)$$

The last equalities mean that $f_2(z) \in B(A_0, A_1)$. On the other hand, f_2 is at least $(p+1)$-valent already in the disk $|z - z_0| < r_p$, which is impossible. We have arrived at the contradiction, and so P_0 lies on the boundary of A. The proof is complete.

CHAPTER II. EXTREMAL PROBLEMS IN THE CLASSES H_p WITH $p < 1$

§1. The form of extremal functions in the classes H_p, $p < 1$

Since the classes H_p, $p < 1$, are not normed (and not even locally convex topological) spaces, the duality method cannot be extended to extremal problems in those classes. Nevertheless, the form of extremal functions in those classes remains the same as for $p \geqslant 1$. Suppose that the domain G is the unit disk. We will start with the Carathéodory-Fejér extremal problem. Let there be given points $\beta_1, \dots, \beta_m$ in the unit disk and the numbers $c_1, \dots, c_m$. Among the functions $f \in H_p$ satisfying the interpolation conditions

$$f(\beta_j) = c_j, \qquad j = 1, \dots, m, \tag{1.1}$$

we must find one for which the infimum

$$\inf \|f\|_p = \inf \left\{ \int_0^{2\pi} |f(e^{i\theta})|^p \, d\theta \right\}^{1/p} \tag{1.2}$$

is attained. Note that we do not exclude the case where the points β_j are multiple. It is clear that the extremal function does exist. As $p \geqslant 1$, it follows from the results in §§1 and 2 of Chapter I that the extremal function $f^*(z)$ has the form

$$f^*(z) = C\prod_{j=1}^{k}\frac{z-\alpha_j}{1-\bar{\alpha}_j z}\prod_{j=1}^{m-1}\left(1-\bar{\alpha}_j z\right)^{2/p}\prod_{j=1}^{m}\left(1-\bar{\beta}_j z\right)^{-2/p}. \tag{1.3}$$

Here C is a constant, the numbers $|\alpha_j| \leqslant 1$, $j = 1,\dots,m-1$, $k \leqslant m-1$, and the first product in (1.3) is taken over those α_j for which $|\alpha_j| < 1$. The extremal function is unique and each $f^*(z)$ of that form is extremal for some problem (1.1).

The form of the function of minimal norm remains the same for $p < 1$.

THEOREM 1.1. *For any $p > 0$ the function $f^*(z)$ of minimal norm has the form* (1.3).

The proof is based on a lemma which will also provide us some additional information later. Let $f^*(z)$ solve the problem (1.1), (1.2) and let $b(z)$ be its Blaschke product. Fix a number $p' > p$ and consider the function

$$g^*(z) = \left[f^*(z)/b(z)\right]^{p/p'}\cdot b(z); \tag{1.4}$$

$g^*(z)$ belongs to $H_{p'}$ and interpolates certain values c_j' at the points β_j.

LEMMA 1.2. *If $f^*(z)$ solves problem* (1.1), (1.2), *then $g^*(z)$ has minimal norm in problem* (1.1), (1.2) *where p is replaced by p' and c_j by c_j'*, $j = 1,\dots,m$.

PROOF. Assume that $g^*(z)$ is not of minimal norm in $H_{p'}$. Then there exists $g(z) \not\equiv g^*(z)$ in $H_{p'}$ interpolating the same values c_j' at the points β_j and such that

$$\|g\|_{p'} \leqslant \|g^*\|_{p'}. \tag{1.5}$$

Construct the function

$$f(z) = \left[g^*(z)/b(z)\right]^{p'/p-1}\cdot g(z). \tag{1.6}$$

From (1.4) it follows that

$$f^*(z) = \left[g^*(z)/b(z)\right]^{p'/p-1}\cdot g^*(z), \tag{1.7}$$

so f^* and f take the same values at the points β_j. Using Hölder's inequality, we obtain

$$\begin{aligned}\left(\|f\|_p\right)^p &= \int_0^{2\pi}\left|f(e^{i\theta})\right|^p d\theta\\ &\leqslant \left(\int_0^{2\pi}\left|f^*(e^{i\theta})\right|^{p'}d\theta\right)^{(p'-p)/p'}\left(\int_0^{2\pi}\left|g(e^{i\theta})\right|^{p'}d\theta\right)^{p/p'}\\ &\leqslant \left(\|g^*\|_{p'}\right)^{p'-p}\cdot\left(\|g\|_{p'}\right)^p \leqslant \left(\|g^*\|_{p'}\right)^{p'} = \left(\|f^*\|_p\right)^p.\end{aligned} \tag{1.8}$$

Since $f^*(z)$ has minimal norm, $\|f\|_p = \|f^*\|_p$. Therefore, $\|g\|_{p'} = \|g^*\|_{p'}$. So, g is the minimal function in the corresponding class.

PROOF OF THE THEOREM. Take $p' > 1$. Then $g^*(z)$ has the form (1.3), where p is replaced by p'. Taking into account the fact that f^* and g^* have the same zeros, from (1.7) we conclude that f^* has the form (1.3).

LEMMA 1.3. *The function* (1.4) *is the unique function in* $H_{p'}$, $p' > p$, *with minimal norm.*

PROOF. Assume that there exists another function $g(z)$ in $H_{p'}$ with minimal norm. Doing the calculation (1.8) over again, we see that $|g(e^{i\theta})| = |g^*(e^{i\theta})|$ almost everywhere. Denote

$$G^*(z) = g^*(z)/b(z), \qquad G(z) = g(z)/b_1(z), \tag{1.9}$$

where $b_1(z)$ is the Blaschke product of $g(z)$. Since $g(z)$ and $g^*(z)$ are the minimal functions, by Theorem 1.1 they have the form (1.3). Therefore, $G^*(z)$ and $G(z)$ are outer functions. Since

$$\left|G^*(e^{i\theta})/G(e^{i\theta})\right| = 1, \tag{1.10}$$

we obtain $G^*(z) = e^{i\gamma}G(z)$, where γ is a real constant. Moreover, since $g^*(\beta_j) = g(\beta_j)$, we have $e^{i\gamma}b(\beta_j) = b_1(\beta_j)$, $j = 1, \ldots, m$. But according to Theorem 1.6 in Chapter I (or Theorem 2.2 in Chapter I), a finite Blaschke product interpolating given values at given points is the minimal function in H_∞, and therefore, it is unique. Thus, $e^{i\gamma}b(z) \equiv b_1(z)$ and, hence, $g^*(z) \equiv g(z)$.

COROLLARY 1.4. *If the function* (1.3) *with given* α_j *and* β_j *is the minimal function for some* $p = p_1$, *then it is also the unique minimal function for all* $p > p_1$ (*with respect to its values at points* β_j).

We call the greatest lower bound of the set of p, for which (1.3) with fixed α_j and β_j is the function with minimal norm, the *minimality index*. The greatest lower bound of those p for which (1.3) is the unique minimal function is called the *uniqueness index*. From Lemmas 1.2 and 1.3 it follows that these indices coincide. If $p_0 > 0$ is the minimality index for (1.3), then an easy continuity argument shows that for $p = p_0$ (1.3) remains the minimal function.

COROLLARY 1.5. *If* $p_0 > 0$ *is the minimality index for the function* (1.3) *with fixed* α_j *and* β_j, *then this function is minimal for* $p \geqslant p_0$ *and is the unique minimal function for* $p > p_0$.

Consideration of the problem (1.1), (1.2) with conditions (1.1) in the form $f(0) = 1$, $f'(0) = c > 0$ shows that for $p = p_0$ uniqueness of functions with minimal norm might fail. We shall not discuss this in greater detail.

Define a mapping of H_p^1 into m-dimensional complex space $\mathbf{C}^m$ by the functionals

$$f(\beta_1), \ldots, f(\beta_m). \tag{1.11}$$

We allow the points β_j to coincide. In that case derivatives of the corresponding orders appear naturally in (1.11). Let A_m be the image of H_p^1 under the mapping (1.11).

THEOREM 1.6. A_m *is a closed bounded set in* $\mathbf{C}^m$ *which contains the origin as an interior point. To every boundary point of* A_m *may correspond only functions* $f^*(z)$ *of the form* (1.3) *with* $\|f^*\|_p = 1$.

PROOF. Closedness and boundedness of A_m follow from compactness of H_p^1. For all sufficiently small c_j, $j = 1, \dots, m$, we can construct a Lagrange interpolation polynomial (or Hermite polynomial in case of multiple points β_j) $R(z)$ which interpolates these values and is such that $\|R\|_p < 1$. This shows that $(0, \dots, 0)$ is an interior point of A_m. Let $(c_1^*, \dots, c_m^*)$ be a boundary point of A_m, and let $f^*(z)$ be one of the functions corresponding to this point. $\|f^*\|_p < 1$ is impossible, because then, using interpolation polynomials again, we could show that $(c_1^*, \dots, c_m^*)$ is an interior point. So, $\|f^*\|_p = 1$ and f^* solves problem (1.1), (1.2) with the values $c_1^*, \dots, c_m^*$. Hence, $f^*(z)$ has the form (1.3).

COROLLARY 1.7. *Let $u(c_1, \dots, c_m)$ be a real-valued function in A_m for which $\sup u(c_1, \dots, c_m)$ can be attained only at the boundary points of A_m. Then only functions of the form* (1.3) *can be extremal in the problem*

$$\sup u(f(\beta_1), \dots, f(\beta_m)), \qquad f \in H_p^1. \tag{1.12}$$

The function u can be, for example, subharmonic. In particular, the assertion 1.7 holds for the linear extremal problem

$$\sup \left| \sum_{j=1}^{m} \gamma_j f(\beta_j) \right| \tag{1.13}$$

with given γ_j.

§2. Extremal problems for nonvanishing functions in H_p, $p > 0$

For each $p > 0$ denote by H_p^0 the subclass of H_p^1 consisting of functions $f(z)$ which do not vanish in the interior of the unit disk. Also, assume that

$$f(0) > 0. \tag{2.1}$$

Thus

$$H_p^0 = \left\{ f \in H_p^1 \colon f(z) \neq 0,\ |z| < 1,\ f(0) > 0 \right\}. \tag{2.2}$$

According to the Smirnov parametric representation, $f(z) \in H_p^0$ has the form

$$f(z) = \exp q(z), \tag{2.3}$$

where

$$q(z) = \frac{1}{2\pi} \int_0^{2\pi} \frac{e^{i\theta} + z}{e^{i\theta} - z} (\ln \rho(\theta)\, d\theta + d\mu(\theta)). \tag{2.4}$$

Here $\rho(\theta) = \ln |f(e^{i\theta})|$, and $d\mu$ is a nonpositive singular Baire measure. Since $f(z) \in H_p^1$, we obtain the normalization condition

$$\int_0^{2\pi} (\rho(\theta))^p\, d\theta \leqslant 1. \tag{2.5}$$

Moreover, the function $\ln \rho(\theta)$ is integrable: $\ln \rho(\theta) \in L_1$.

Setting $\rho(\theta) = [r(\theta)]^{1/p}$, we obtain the following representation for $q(z)$:

$$q(z) = \frac{1}{2\pi p} \int_0^{2\pi} \frac{e^{i\theta} + z}{e^{i\theta} - z} (\ln r(\theta)\, d\theta + d\mu(\theta)), \tag{2.6}$$

where $r(\theta) \geqslant 0$ and

$$\int_0^{2\pi} r(\theta)\, d\theta \leqslant 1, \qquad d\mu \leqslant 0 \text{ singular}. \tag{2.7}$$

Conversely, for any such $q(z)$, the function $f(z) = \exp q(z) \in H_q^0$.

Let us introduce the following classes of functions: Q_p^* consists of functions $q(z)$ of the form (2.6) satisfying (2.7), and Q_p consists of functions of the form

$$q(z) = \frac{1}{2\pi p} \int_0^{2\pi} \frac{e^{i\theta} + z}{e^{i\theta} - z} \ln r(e)\, d\theta \tag{2.8}$$

satisfying the first of the conditions (2.7). So, Q_p^* consists of logarithms of H_p^0-functions.

Let us consider also the class σ of integrable functions $S(\theta)$ on $[0, 2\pi]$, i.e., $S(\theta) \in L_1$ such that

$$\int_0^{2\pi} e^{S(\theta)}\, d\theta \leqslant 1, \tag{2.9}$$

and the class Σ of Baire measures ν of the form

$$d\nu = S(\theta)\, d\theta + d\mu, \tag{2.10}$$

where $S(\theta) \in \sigma$ and $d\mu \leqslant 0$ is a singular measure. For these classes of functions and measures we have the obvious relations $Q_p \subset Q_p^*$ and $\sigma \subset \Sigma$.

LEMMA 2.1. *In order that $q(z) \in Q_p$, it is necessary and sufficient that*

$$q(z) = \frac{1}{2\pi p} \int_0^{2\pi} \frac{e^{i\theta} + z}{e^{i\theta} - z} S(\theta)\, d\theta, \qquad S(\theta) \in \sigma. \tag{2.11}$$

In order that $q(z) \in Q_p^*$, it is necessary and sufficient that

$$q(z) = \frac{1}{2\pi p} \int_0^{2\pi} \frac{e^{i\theta} + z}{e^{i\theta} - z}\, d\nu, \qquad \nu \in \Sigma. \tag{2.12}$$

In order that $f(z) \in H_p^0$, it is necessary and sufficient that (2.3) *hold with $q(z) \in Q_p^*$.*

LEMMA 2.2. *The classes Q_p and Q_p^* are convex.*

PROOF. It suffices to prove the assertion for the class Q_p. Let $q_i(z) \in Q_p$, $i = 1, 2$, and $S_i(\theta)$, $i = 1, 2$, be functions in σ that correspond to $q_i(z)$ via (2.11). Then

$$\int_0^{2\pi} e^{S_i(\theta)} \leqslant 1, \qquad i = 1, 2.$$

If $0 < t < 1$, then, since the exponential function is convex,

$$e^{tS_1(\theta) + (1-t)S_2(\theta)} \leqslant te^{S_1(\theta)} + (1-t)e^{S_2(\theta)}.$$

This immediately implies that (2.9) holds for $S(\theta) = tS_1(\theta) + (1-t)S_2(\theta)$. Moreover, $S(\theta)$ is integrable, and so $S(\theta) \in \sigma$.

LEMMA 2.3. *Let $\alpha(\theta)$ be a continuous function on* $[0, 2\pi]$. *For the supremum*

$$\sup_{S(\theta)\in\sigma} \int_0^{2\pi} S(\theta)\alpha(\theta)\, d\theta \tag{2.13}$$

to be finite, it is necessary and sufficient that $\alpha(\theta) \geqslant 0$.

PROOF. *Necessity*. Suppose that $\alpha(\theta) \leqslant -\varepsilon < 0$ on some interval $l \subset [0, 2\pi]$. Take an arbitrarily large $N > 0$ and $k > 0$ such that $2\pi e^{-k} < \frac{1}{2}$. Construct a function $S_N(\theta)$ defined by

$$S_N(\theta) = \begin{cases} -N, & \theta \in l, \\ -k, & \theta \in [0, 2\pi] \setminus l. \end{cases} \tag{2.14}$$

It is easy to check that $S_N(\theta) \in \sigma$ for all sufficiently large N. At the same time,

$$\int_0^{2\pi} S_N(\theta)\alpha(\theta)\, d\theta \geqslant lN\varepsilon_0 - k \max|\alpha(\theta)| \cdot 2\pi. \tag{2.15}$$

For $N \to +\infty$ and k fixed the latter quantity tends to $+\infty$.

Sufficiency. If $\alpha(\theta) \geqslant 0$, then

$$\int_0^{2\pi} S(\theta)\alpha(\theta)\, d\theta \leqslant \int_0^{2\pi} e^{S(\theta)}\alpha(\theta)\, d\theta \leqslant \max|\alpha(\theta)| \leqslant +\infty. \tag{2.16}$$

LEMMA 2.4. *Let $\alpha(\theta)$ be a continuous function on* $[0, 2\pi]$. *For the supremum*

$$\sup_{\nu\in\Sigma} \int_0^{2\pi} \alpha(\theta)\, d\nu \tag{2.17}$$

to be finite, it is necessary and sufficient that $\alpha(\theta) \geqslant 0$.

PROOF. Since $\sigma \subset \Sigma$, necessity follows from Lemma 2.3. Sufficiency follows from the fact that $d\mu \leqslant 0$ in (2.10).

LEMMA 2.5. *Let $\alpha(\theta) \geqslant 0$ be continuous on* $[0, 2\pi]$. *Then*

$$\sup_{S(\theta)\in\sigma} \int_0^{2\pi} S(\theta)\alpha(\theta)\, d\theta = \sup_{\nu\in\Sigma} \int_0^{2\pi} \alpha(\theta)\, d\nu = \int_0^{2\pi} \ln \frac{\alpha(\theta)}{A} \cdot \alpha(\theta)\, d\theta, \tag{2.18}$$

where

$$A = \int_0^{2\pi} \alpha(\theta)\, d\theta. \tag{2.19}$$

If $\ln \alpha(\theta)$ *is integrable, then the supremum in* (2.18) *for the class σ is attained only for the functions*

$$S^*(\theta) = \ln \frac{\alpha(\theta)}{A}, \tag{2.20}$$

and the supremum in Σ is attained only for the measures

$$d\nu^* = S^*(\theta)\, d\theta + d\mu^*, \tag{2.21}$$

where $d\mu^* \leqslant 0$ is a singular measure whose support S_{μ^*} is contained in the zero set of $\alpha(\theta)$.

If $\ln \alpha(\theta) \notin L_1$, *then the suprema* (2.18) *in the classes σ and Σ are not attained.*

PROOF. It clearly suffices to consider those σ for which equality holds in (2.9). Note that for any positive u and v we have

$$u \ln u - u \geqslant u \ln v - v, \tag{2.22}$$

where equality obtains only if $v = u$. The proof of (2.22) is a standard exercise in differential calculus. In (2.22) set $u = \alpha(\theta)/A$ and $\ln v = S(\theta)$, and integrate

$$\int_0^{2\pi} \frac{\alpha(\theta)}{A} \ln \frac{\alpha(\theta)}{A}\, d\theta - \int_0^{2\pi} \frac{\alpha(\theta)}{A}\, d\theta \geqslant \int_0^{2\pi} \frac{\alpha(\theta)}{A} S(\theta)\, d\theta - \int_0^{2\pi} e^{S(\theta)}\, d\theta.$$

Since $\int_0^{2\pi} (\alpha(\theta)/A)\, d\theta = \int_0^{2\pi} e^{S(\theta)}\, d\theta = 1$, we see that

$$\int_0^{2\pi} \ln \frac{\alpha(\theta)}{A} \alpha(\theta)\, d\theta \geqslant \int_0^{2\pi} S(\theta)\alpha(\theta)\, d\theta. \tag{2.23}$$

If $\ln(\alpha(\theta)/A) \in L_1$, then $\ln(\alpha(\theta)/A) \in \sigma$ and the above argument implies that equality in (2.23) holds only if $S(\theta) = \ln(\alpha(\theta)/A)$. If $\ln(\alpha(\theta)/A) \notin L_1$, and, therefore, $\ln(\alpha(\theta)/A) \notin \sigma$, using a simple approximation technique we verify that

$$\sup_{S(\theta) \in \sigma} \int_0^{2\pi} S(\theta)\alpha(\theta)\, d\theta = \int_0^{2\pi} \ln \frac{\alpha(\theta)}{A} \cdot \alpha(\theta)\, d\theta. \tag{2.24}$$

But now the supremum cannot be attained in the class σ. Since $\sigma \subset \Sigma$ and, on the other hand, $d\mu \leqslant 0$ in (2.10), we have

$$\sup_{\nu \in \Sigma} \int_0^{2\pi} \alpha(\theta)\, d\nu = \sup_{S(\theta) \in \sigma} \int_0^{2\pi} S(\theta)\alpha(\theta)\, d\theta. \tag{2.25}$$

If $d\nu^* = S^*(\theta)\, d\theta + d\mu^*$ is an extremal measure in Σ, then

$$\int_0^{2\pi} \alpha(\theta)\, d\nu^* = \int_0^{2\pi} \alpha(\theta)[S^*(\theta)\, d\theta + d\mu^*] = \int_0^{2\pi} \alpha(\theta) \ln \frac{\alpha(\theta)}{A}\, d\theta.$$

Therefore, $S^*(\theta)$ has the form (2.20). Also, $d\mu^*$ must be concentrated on the set where $\alpha(\theta) = 0$, in order not to diminish the integral. The lemma is proved.

These lemmas allow us to investigate extremal problems in the classes Q_p and Q_p^*. Let a function $q(z)$ have the decomposition

$$q(z) = \sum_{k=0}^{\infty} c_k z^k. \tag{2.26}$$

Let $\gamma_0, \gamma_1, \dots, \gamma_{m-1}$ be given complex numbers. Consider the extremal problems

$$\lambda_p = \sup_{q \in Q_p} \operatorname{Re} \sum_{k=0}^{m-1} \gamma_k c_k, \qquad \lambda_p^* = \sup_{q \in Q_p^*} \operatorname{Re} \sum_{k=0}^{m-1} \gamma_k c_k. \tag{2.27}$$

THEOREM 2.6. *Let*

$$\alpha(\theta) = \operatorname{Re}\left(\gamma_0 + 2 \sum_{k=1}^{m-1} \gamma_k e^{-ik\theta} \right). \tag{2.28}$$

For λ_p and λ_p^* to be finite it is necessary and sufficient that $\alpha(\theta) \geqslant 0$ on $[0, 2\pi]$. In the latter case

$$\lambda_p = \lambda_p^* = \frac{1}{2\pi p} \int_0^{2\pi} \ln \frac{\alpha(\theta)}{A} \cdot \alpha(\theta)\, d\theta, \qquad A = \int_0^{2\pi} \alpha(\theta)\, d\theta. \tag{2.29}$$

The unique extremal function in the class Q_p has the form

$$q(z) = \frac{2}{p} \sum_{k=1}^{m-1} \ln(1 - \bar{\alpha}_k z) + \ln \frac{a}{A}. \tag{2.30}$$

Here the α_k, $|\alpha_k| \leqslant 1$, are points in the disk $|z| \leqslant 1$ and $a > 0$ is a constant such that

$$\alpha(\theta) = a \prod_{k=1}^{m-1} \left|1 - \bar{\alpha}_k e^{i\theta}\right|^2. \tag{2.31}$$

Extremal functions $q^*(z)$ in the class Q_p^* are given by

$$q^*(z) = q(z) + \sum{}' \lambda_k \frac{\alpha_k + z}{\alpha_k - z}, \tag{2.32}$$

where Σ' is taken over those points α_k for which $|\alpha_k| = 1$. The λ_k's are arbitrary nonpositive numbers.

Proof. The finiteness of λ_p and λ_p^* follows from Lemma 2.3. Since for $q(z) \in Q_p$ (2.11) and (2.12) imply that

$$\operatorname{Re} \sum_{k=0}^{m-1} \gamma_k c_k = \frac{1}{2\pi p} \int_0^{2\pi} \alpha(\theta) S(\theta)\, d\theta,$$

and a similar formula with $S(\theta)\, d\theta$ replaced by $d\nu$ holds for $q(z) \in Q_p^*$, (2.29) follows from (2.18). Since $\alpha(\theta)$ is a nonnegative trigonometric polynomial of order $m-1$, there exist points $\alpha_1, \ldots, \alpha_m$ in $|z| \leqslant 1$ and a number $a > 0$ such that (2.31) holds. The latter fact is the well-known Fejér-Riesz theorem (see, for instance, [4], Russian p. 328 or [33], Chapter III, Theorem 2.1) on the representation of nonnegative trigonometric polynomials. Taking the extremal measure $S(\theta) d\theta = \ln(\alpha(\theta)/A)\, d\theta$ in (2.11), we obtain (2.30). (2.32) follows from Lemma 2.5 (formula (2.21)) and (2.12).

Let us give an example of the concrete estimate obtained from this theorem. Consider the problem of

$$\sup_{q \in Q_p^*} \operatorname{Re}(c_m + 2c_0), \qquad m = 1, 2, \ldots. \tag{2.33}$$

We have

$$\begin{aligned} \sup \operatorname{Re}(c_m + 2c_0) &= \sup \frac{1}{\pi p} \int_0^{2\pi} (1 + \cos m\theta)\, d\nu \\ &= \frac{1}{\pi p} \int_0^{2\pi} (1 + \cos m\theta) \ln \frac{(1 + \cos m\theta)}{2\pi}\, d\theta \\ &= \frac{1}{\pi p} \int_0^{2\pi} (1 + \cos \theta) \ln \frac{(1 + \cos \theta)}{2\pi}\, d\theta. \end{aligned} \tag{2.34}$$

Thus, the extremum problem in (2.33) does not depend on m. But extremal functions in Q_p and Q_p^* for problem (2.33) do depend on m. We leave it to the reader to write down the corresponding expressions.

Consider the sets $A_m(Q_p)$ and $A_m(Q_p^*)$ in the real $(2m-1)$-dimensional space R^{2m-1} generated by the first m coefficients $(c_0, c_1, \ldots, c_{m-1})$ of the functions $q(z)$ in the classes Q_p and Q_p^* respectively. We replace each complex coordinate c_j, $j = 1, \ldots, m-1$, as usual, by the pair of real coordinates $\operatorname{Re} c_j$, $\operatorname{Im} c_j$. The coordinate c_0 is always real; therefore, we must consider the sets $A_m(Q_p)$ and $A_m(Q_p^*)$ in R^{2m-1}.

THEOREM 2.7. *$A_m(Q_p)$ and $A_m(Q_p^*)$ are convex sets with nonempty interior. The set $A_m(Q_p^*)$ is closed. The closure of $A_m(Q_p)$ coincides with $A(Q_p^*)$. Both sets are unbounded, and $c_0 \to -\infty$ as $(c_0, \ldots, c_{m-1}) \to \infty$. To each boundary point of $A_m(Q_p^*)$ there corresponds a unique function $q(z)$ of the form* (2.32).

PROOF. Convexity of $A_m(Q_p)$ and $A_m(Q_p^*)$ follows from the convexity of Q_p and Q_p^*. Let us show that $A_m(Q_p)$ has a nonempty interior; then $A_m(Q_p^*)$ also has interior points. Assume this is false. Then there exists a hyperplane π: $\operatorname{Re}\sum_0^{m-1} \gamma_k c_k = d$ which contains $A_m(Q_p)$. Here $\gamma_0, \ldots, \gamma_{m-1}$ are complex coefficients (we can assume γ_0 to be real). $c_0, \ldots, c_{m-1}$ are coordinates of the current point. Since π contains $A_m(Q_p)$, it follows that

$$\operatorname{Re} \int_0^{2\pi} \left(\gamma_0 + 2 \sum_{k=1}^{m-1} \gamma_k e^{-ik\theta} \right) S(\theta)\, d\theta = d, \tag{2.35}$$

for all $S(\theta) \in \sigma$. According to Lemma 2.3, this implies that $\alpha(\theta) = \operatorname{Re}(\gamma_0 + \sum_1^{m-1} \gamma_k e^{-ik\theta})$ on $[0, 2\pi]$. If there existed a θ_0 such that $\alpha(\theta) > 0$ in a neighborhood of θ_0, then (2.35) would not hold for all $S(\theta) \in \sigma$. In fact, taking $S(\theta) \in \sigma$, change it in the neighborhood of θ_0, setting $S_1(\theta) = S(\theta) - \delta$ in this neighborhood, $\delta > 0$, and $S_1(\theta) = S(\theta)$ elsewhere. It is clear that $S_1(\theta) \in \sigma$; but

$$\int_0^{2\pi} \alpha(\theta) S_1(\theta)\, d\theta < \int_0^{2\pi} \alpha(\theta) S(\theta)\, d\theta,$$

and so (2.35) does not hold for the whole class σ. Thus, $\alpha(\theta) \equiv 0$. From this, taking into account the structure of $\alpha(\theta)$, one can easily verify that $\gamma_0 = \gamma_1 = \cdots = \gamma_{m-1} = 0$, which cannot happen in the equation of the hyperplane. So we have a contradiction, and therefore, $A_m(Q_p)$ has interior points.

Now let us show that $A_m(Q_p^*)$ is closed. Let the sequence $(c_0^n, \ldots, c_{m-1}^n)$ in $A_m(Q_p^*)$ converge to a point $(c_0, \ldots, c_{m-1})$ as $n \to \infty$. Denote by $q_n(z)$ the function in Q_p^* corresponding to the point $(c_0^n, \ldots, c_{m-1}^n)$. The functions $f_n(z) = \exp q_n(z)$ belong to H_p^0, and we can assume that they converge uniformly on compact subsets of the disk $|z| < 1$ either to $f(z) \in H_p^0$ or to $f(z) \equiv 0$. The latter cannot occur, since this would imply that $\operatorname{Re} q_n(z) \to -\infty$, whereas we have $q_n(0) = c_0^n \to c_0$. So, $f(z) \in H_p^0$, and therefore, $f(z) = \exp q(z)$, $q(z) \in Q_p^*$. It is easy to see that $q(z)$ has $c_0, \ldots, c_{m-1}$ as its first Taylor coefficients. Hence, $(c_0, \ldots, c_{m-1}) \in A_m(Q_p^*)$. Moreover, from the above argument it also follows that the sets $A_m(Q_p)$ and $A_m(Q_p^*)$ are unbounded.

Now let $(c_0^*, \ldots, c_{m-1}^*)$ be a boundary point of $A_m(Q_p^*)$. Draw a supporting hyperplane to $A_m(Q_p^*)$ passing through this point:

$$\operatorname{Re} \sum_{k=0}^{m-1} \gamma_k c_k = d. \tag{2.36}$$

For all points of $A_m(Q_p^*)$ we can assume that

$$\operatorname{Re} \sum_{k=0}^{m-1} \gamma_k c_k \leqslant d, \tag{2.37}$$

which becomes an equality for $(c_0^*, \ldots, c_{m-1}^*)$. Using (2.37) we obtain that the supremum

$$d = \sup_{\nu \in \Sigma} \int_0^{2\pi} \left(\gamma_0 + 2 \operatorname{Re} \sum_{k=1}^{m-1} \gamma_k e^{-ik\theta} \right) d\nu = \sup_{\nu \in \Sigma} \int_0^{2\pi} \alpha(\theta)\, d\nu \tag{2.38}$$

is finite and attained for the measures ν^* corresponding to the point $(c_0^*, \ldots, c_{m-1}^*)$. From this, according to Theorem 2.6, it follows that the function $q^*(z)$ corresponding to this point has the form (2.32). Let us prove its uniqueness, i.e., uniqueness of the measure $\nu^* \in \Sigma$ corresponding to our point. Since ν^* is extremal in (2.38), then according to Lemma 2.5 its absolutely continuous part $\ln(\alpha(\theta)/A)$ is uniquely defined. We must show that the atoms μ_k of this measure, situated at the zeros θ_k of the polynomial $\alpha(\theta)$, are also uniquely defined. The number of those zeros does not exceed the degree $m-1$ of $\alpha(\theta)$, $k = 1, \ldots, r \leqslant m-1$. Let μ be a measure which consists of atoms μ_k at the points θ_k, $k = 1, \ldots, r$, i.e., μ is the singular part of measure ν^* corresponding to the boundary point under consideration. Let

$$C_0 = \int_0^{2\pi} d\mu, \qquad C_k = \frac{1}{2} \int_0^{2\pi} e^{-ik\theta}\, d\mu, \quad k = 1, \ldots, m-1. \tag{2.39}$$

Since

$$C_0 = c_0^* - \int_0^{2\pi} \ln \frac{\alpha(\theta)}{A}\, d\theta, \qquad C_k = c_k^* - \frac{1}{2} \int_0^{2\pi} e^{ik\theta} \ln \frac{\alpha(\theta)}{A}\, d\theta, \tag{2.40}$$

the numbers C_k, $k = 0, 1, \ldots, m-1$, are uniquely determined by the point $(c_0^*, \ldots, c_{m-1}^*)$. For any polynomial $R(\theta) = \gamma_0 + \sum_1^{m-1} \gamma_k e^{-ik\theta}$ we have

$$\int_0^{2\pi} R(\theta)\, d\mu = \gamma_0 c_0 + 2 \sum_{k=1}^{m-1} \gamma_k c_k. \tag{2.41}$$

For a fixed zero θ_j, let us construct a polynomial $R_j(\theta)$ of the above form which interpolates the conditions

$$R_j(\theta_j) = 1, \qquad R_j(\theta_k) = 0, \quad k = 1, \ldots, r,\ k \neq j. \tag{2.42}$$

Then we have

$$\int_0^{2\pi} R_j(\theta)\, d\mu = \mu_j = \sum_{k=0}^{m-1} C_k \gamma_k^j, \tag{2.43}$$

where the γ_k^j are the coefficients of $R_j(\theta)$. From (2.43) it follows that μ_j is uniquely determined.

Finally, coming back to the extremal problem (2.38), we recall that $\ln(\alpha(\theta)/A)$ is also a solution of it (Lemma 2.5). Therefore, the hyperplane (2.36) contains points of $A_m(Q_p)$. This immediately implies that the closure of $A_m(Q_p)$ coincides with $A_m(Q_p^*)$. All assertions of the theorem are proved.

Let us go back to the class H_p^0. For this class, let us consider the set $A_m(H_p^0)$ constructed similarly to $A_m(Q_p)$ and $A_m(Q_p^*)$. It is clear that the point $(0) = (0, \ldots, 0) \notin A_m(H_p^0)$. Set $\overline{A}_m(H_p^0) = A_m(H_p^0) \cup (0)$.

THEOREM 2.8. *$\overline{A_m(H_p^0)}$ is a bounded closed set with a nonempty interior (with respect to R^{2m-1}). To boundary points of $\overline{A_m(H_p^0)}$ different from (0) correspond only functions of the form*

$$f^*(z) = C \prod_{k=1}^{m-1} (1 - \bar{\alpha}_k z)^{2/p} \cdot \exp \sum{}' \lambda_k \frac{\alpha_k + z}{\alpha_k - z}, \tag{2.44}$$

where the α_k, λ_k and $\sum'$ are the same as in (2.32) and $C > 0$ is such that

$$\int_0^{2\pi} |f^*(e^{i\theta})|^p \, d\theta = 1. \tag{2.45}$$

PROOF. It is clear that $\overline{A_m(H_p^0)}$ is bounded and closed. Add to $A_m(Q_p)$ the element ∞ by defining that $(c_0, \ldots, c_{m-1}) \in A_m(Q_p)$ tends to ∞ as $c_0 \to -\infty$. Define a mapping of $A_m(Q_p^*)$ onto $A_m(H_p^0)$ by (2.3). It is not hard to check that this mapping is a homeomorphism. Hence $\overline{A}_m(H_p^0)$ contains interior points and each boundary point of $\overline{A}_m(H_p^0)$ different from (0) must be an image of a boundary point of $A_m(Q_p^*)$. Now it remains to refer to Theorem 2.7.

Since now we know what functions correspond to the boundary points of $\overline{A_m(H_p^0)}$, we can conclude that extremal functions in that class in the problems of the type $\sup|\sum_0^{m-1} \gamma_k c_k|$ and in more general problems have the form (2.44).

We could also carry out this presentation for more general functionals than Taylor coefficients, e.g., linear combinations of the derivatives taken at several points, etc.

CHAPTER III. EXTREMAL PROBLEMS FOR BOUNDED ANALYTIC FUNCTIONS SATISFYING ADDITIONAL CONDITIONS IN THE INTERIOR OF THE DOMAIN

§1. Posing of extremal problems with additional conditions. Milloux's theorem

In this chapter we present a general approach to the study of extremal problems for classes of analytic functions which satisfy a certain restriction on growth in their domains of definition, and some additional restriction on growth on a subset of the domain. A typical example is given by the problem whose study was initiated by Milloux in [39].

Consider the class $B^{1,\varepsilon}$ of functions $f(z)$ analytic in the disk $|z| < 1$, bounded there by 1, and satisfying the following condition. For each $f(z) \in B^{1,\varepsilon}$ there exists a set $\mathscr{D}$ (depending on the function) which has nonempty intersection with every circle $|z| = R$, $0 < R < 1$, and such that $|f(z)| \leqslant \varepsilon < 1$ for $z \in \mathscr{D}$. The problem is to find or to estimate

$$\sup_{f \in B^{1,\varepsilon}} \max_{|z|=r} |f(z)|.$$

The following theorem is due to Milloux [39].

THEOREM 1.1. *Functions $f(z) \in B^{1,\varepsilon}$ satisfy the estimate*

$$|f(z)| \leqslant \varepsilon^{C[1-|z|]}, \tag{1.1}$$

where C is an absolute constant.

We shall give a proof of this fact, and we shall see that we can take $1/\pi$ for C.

PROOF. It obviously suffices to consider the subclass $B_{+}^{1;\varepsilon}$ of $B^{1,\varepsilon}$ consisting of functions such that

$$\max_{|z|=r} |f(z)| = f(r).$$

Let $f(z) \in B_{+}^{1;\varepsilon}$. Since $|f(z)| \leqslant 1$, we have

$$f(z) = \prod_{i=1}^{\infty} \frac{\alpha_i - z}{1 - \bar{\alpha}_i z} \frac{|\alpha_i|}{\alpha_i} \exp\left(\frac{1}{2\pi} \int_0^{2\pi} \frac{e^{i\theta} + z}{e^{i\theta} - z}\, d\mu(\theta) \right),$$

where $\{\alpha_i\}$ are all the zeros of $f(z)$ and $\mu(\theta)$ is a nonincreasing function. (If $\alpha_i = 0$, then the factor $|\alpha_i|/\alpha_i$ is not introduced.)

We need the estimate

$$\left| \frac{-|z| + \alpha}{1 - |\alpha|\,|z|} \right| \leqslant \left| \frac{z - \alpha}{1 - \bar{\alpha} z} \right| \leqslant \frac{|z| + |\alpha|}{1 + |z|\,|\alpha|}, \tag{1.2}$$

which is an immediate corollary of the identity

$$1 - \left| \frac{z - \alpha}{1 - \bar{\alpha} z} \right| = \frac{(1 - |\alpha|^2)(1 - |z|^2)}{|\alpha|\,|1/\bar{\alpha} - z|^2}. \tag{1.3}$$

Consider the function

$$h(z) = \prod_{i=1}^{\infty} \frac{|\alpha_i| + z}{1 + |\alpha_i| z} \cdot \exp\left(\frac{1 - z}{1 + z} \int_0^{2\pi} d\mu \right). \tag{1.4}$$

From (1.2) and the well-known Harnack's inequalities for

$$\frac{1 - z^2}{1 + z^2 - 2r\cos(\theta - \varphi)} = \operatorname{Re} \frac{e^{i\theta} + re^{i\varphi}}{e^{i\theta} - re^{i\varphi}}, \qquad 0 < r < 1,$$

it follows that

$$h(-|z|) \leqslant |f(z)|, \qquad h(|z|) \geqslant f(|z|). \tag{1.5}$$

These inequalities ensure the relation $h(z) \in B_{+}^{1;\varepsilon}$, and for $h(z)$ we have $D = (-1, 0]$.

Thus, the problem has been reduced to the following. (The above reduction is due to Heins.) Find

$$\sup f(z), \qquad 0 < r < 1, \tag{1.6}$$

over all functions $f(z)$, analytic in the disk $|z| < 1$, such that

$$|f(z)| \leqslant 1, \quad |z| < 1, \quad \text{and} \quad |f(z)| \leqslant \varepsilon < 1, \quad z \in (-1, 0]. \tag{1.7}$$

Consider the mapping

$$w = \sqrt{-\tfrac{1}{2}\left(z + \tfrac{1}{2}\right) + 1}$$

of the unit disk slit along $(-1, 0]$ onto the upper half-plane. The function $f(z)$ transforms into the function $F(w)$, $|F(w)| \leqslant 1$. Taking, if necessary, $f(\rho z)$, $0 < \rho < 1$ and letting $\rho \to 1$, we can assume that $f(z)$ is continuous in the closed disk $|z| \leqslant 1$. Then $F(w)$ is continuous up to the real axis and satisfies there the inequalities

$$|F(w)| < 1, \qquad |F(w)| \leqslant \varepsilon, \qquad w \in (\sqrt{2}, +\infty) \cup (-\infty, -\sqrt{2}).$$

The point r is mapped onto the point

$$\sqrt{\tfrac{1}{2}\left(r + \tfrac{1}{2}\right) - 1} \cdot i = ki$$

on the imaginary axis. Majorizing the subharmonic function $\ln|F(w)|$ by a function $u(w)$ harmonic for $\operatorname{Im} w \geqslant 0$ and satisfying the boundary conditions

$$u(w) = 0, \qquad w \in (-\sqrt{2}, \sqrt{2});$$

$$u(w) = \ln \varepsilon, \qquad w \in (-\infty, -\sqrt{2}) \text{ or } w \in (\sqrt{2}, +\infty)$$

and representing $u(w)$ by the Poisson integral in the half-plane, we find that

$$\ln f(r) = \ln F(ki) \leqslant k \ln \varepsilon \cdot \frac{2}{\pi} \int_{\sqrt{2}}^{\infty} \frac{dx}{x^2 + k^2}$$

$$= \ln \varepsilon \cdot \frac{2}{\pi}\left(\frac{\pi}{2} - \arctan \frac{\sqrt{2}}{k}\right).$$

Recalling the value of k and setting $\beta = 1 - r$, we transform this estimate into

$$\ln f(r) \leqslant \ln \varepsilon \cdot \frac{2}{\pi}\left(\frac{\pi}{2} - \arctan \frac{2\sqrt{1-\beta}}{\beta}\right).$$

Consider the function

$$\varphi(\beta) = \frac{\pi}{2} - \arctan \frac{2\sqrt{1-\beta}}{\beta} - \frac{1}{2}\beta, \qquad 0 < \beta < 1.$$

We find that $\varphi'(\beta) > 0$. Hence, $\varphi(\beta) > \varphi(0) = 0$. Therefore,

$$\frac{\pi}{2} - \arctan \frac{2\sqrt{1-\beta}}{\beta} \geqslant \frac{1}{2}\beta.$$

This leads to the estimate

$$\ln f(r) \leqslant \frac{1}{\pi} \ln \varepsilon \cdot \beta = \frac{1}{\pi} \ln \varepsilon (1 - r).$$

Thus, $f(r) \leqslant \varepsilon^{(1-r)/\pi}$. Q.E.D.

We can supplement the above theorem by showing that the value $C = 1/\pi$ is the best possible.

Consider the harmonic function $u(w)$ used in the proof of Milloux's theorem. Let $v(w)$ be its harmonic conjugate. The analytic function $F(w) = e^{u+iv}$ satisfies in the half-plane $\operatorname{Im} w > 0$ the conditions

$$|F(w)| \leqslant 1, \qquad |F(w)| \leqslant \varepsilon, \qquad w \in (-\infty, -\sqrt{2}) \cup (\sqrt{2}, +\infty),$$

$$F(-w) = F(w).$$

The function $f(z)$ obtained from $F(w)$ by pulling it back to the disk $|z| < 1$ slit along $(-1, 0]$ is analytic everywhere in $|z| < 1$, since it has the same values on both sides of the slit. Also, it is clear that $f(z) \in B^{1,\varepsilon}$. But

$$\ln f(r) = \left(\frac{\pi}{2} - \arctan\frac{2\sqrt{1-\beta}}{\beta}\right) \cdot \frac{2}{\pi} \ln \varepsilon.$$

Since

$$\lim_{\beta \to 0\,(r \to 1)} \frac{\pi/2 - \arctan\left(2\sqrt{1-\beta}/\beta\right)}{\beta} = \frac{1}{2},$$

the inequality $\ln f(r) \leqslant C \ln \varepsilon \cdot (1 - r)$ fails for $C > 1/\pi$, which is what we needed to show.

In this chapter we present a duality method for investigating extremal problems of Milloux's type. Instead of the functional $f(z_0)$, one can consider other functionals, for instance $f'(z_0)$, etc. Also, instead of $B^{1,\varepsilon}$ one can consider other classes distinguished by different restrictions, e.g., a bound on the growth of integral means. First of all, we shall concentrate on the qualitative study of extremal functions. To illustrate the idea of our approach, let us show how one can interpret the (reduced) Milloux problem.

Let G be a finitely connected domain with a rectifiable boundary Γ. Let $\mathscr{D}$ be a subset of G. Let $X = \tilde{M}(\Gamma)$ be the linear space of bounded measurable functions on Γ with the norm $\|f\| = \operatorname{ess\,sup}|f(\zeta)|$. Take the subspace E in X consisting of boundary values of bounded analytic functions in G. Consider the unit sphere S_1 in X and the set $B = S_1 \cap E$. Thus B is the class of boundary values of functions $f(z)$ analytic in G such that $|f(z)| \leqslant 1$. We also consider the space $X_1 = C(\mathscr{D})$ of continuous functions on $\mathscr{D}$ (with the usual metric). Let us consider the sphere S of radius ε in X_1 and the linear operator

$$Nf = \varphi(z) = \frac{1}{2\pi i}\int_\Gamma \frac{f(\zeta)\,d\zeta}{\zeta - z}, \qquad z \in \mathscr{D},$$

from X into X_1. Our class of functions consists of those elements of B for which $Nf \in S$. Denote this class by B^S. The problem (1.6), (1.7) is a particular case of the problem about the extremum of a linear functional on B^S. In the next section we will extend the duality method to such problems.

§2. Duality relations for problems with additional conditions([1])

THEOREM 2.1. *Let $X = \{f\}$ be a normed space with unit sphere S_1. Let E be a subspace in X, and $B = S_1 \cap E$. Let X_1 be another normed space and let S be the sphere in that space centered at the origin with radius ε. Let Nf be a continuous linear operator from X to X_1. Let B^S be the set of elements in B for which $Nf \in S$, i.e.,*

$$B^S = \{f\colon f \in E,\ \|f\| \leqslant 1,\ Nf \in S\}. \tag{2.1}$$

([1])There are many books and papers that study duality in general extremal problems. For instance, see [18].

Let $l_0(f)$ be a linear functional on X. Then

$$\sup_{f\in B^S} |l_0(f)| = \inf_{\Lambda, m} \left[\|l_0 - \tilde{l} - m\| + \varepsilon\|\Lambda\|\right], \tag{2.2}$$

where Λ runs over the whole space X_1^* dual to X_1 and m runs over the annihilator $E^{\perp} \subset X^*$ of the subspace E. $\tilde{l}$ is the functional on X defined by

$$\tilde{l}(f) = \Lambda(Nf) \qquad (\Lambda \in X_1^*). \tag{2.3}$$

(In other words, $\tilde{l} = N^*\Lambda$, where N^* is the adjoint of N.) The infimum in (2.2) is attained.

PROOF. Consider the direct product $X_0 = X \times X_1$. Elements of this space are pairs $\mathscr{F} = (f, \varphi)$, where $f \in X$ and $\varphi \in X_1$. Set

$$\|\mathscr{F}\| = \max\{\|f\|, \|f\|/\varepsilon\}. \tag{2.4}$$

The dual space X_0^* of X_0 consists of linear functionals L of the form $L = (l, \Lambda)$, where $l \in X^*$, $\Lambda \in X_1^*$, and

$$L(\mathscr{F}) = L(f\varphi) = l(f) - \Lambda(\varphi); \tag{2.5}$$

here

$$\|L\| = \|l\| + \varepsilon\|\Lambda\|. \tag{2.6}$$

Take a subspace $\mathscr{E}$ in X_0 consisting of elements of the form (f, Nf), $f \in E$. For such elements a condition of the form

$$\|\mathscr{F}\| = \|(f, Nf)\| \leqslant 1 \tag{2.7}$$

means that

$$\|f\| \leqslant 1, \qquad \|Nf\| \leqslant \varepsilon. \tag{2.8}$$

Hence, (2.7) is equivalent to the inclusion $f \in B^S$. Let us investigate the structure of the annihilator $\mathscr{E}^{\perp}$ in X_0^*. If a functional $\lambda = (l, \Lambda) \in \mathscr{E}^{\perp}$, then

$$l(f) - \Lambda(Nf) = 0 \qquad \forall f \in E. \tag{2.9}$$

Therefore the functional l in the pair (l, Λ) has the structure

$$l = \tilde{l} + m, \tag{2.10}$$

where $\tilde{l}$ has the form (2.3) and $m \in E^{\perp}$. Conversely, every functional (l, Λ) with l of the form (2.10) belongs to $\mathscr{E}^{\perp}$. Let us consider a functional $L_0 = (l_0, 0)$ on X_0. According to (1.1) in [55], we have

$$\sup_{\substack{\mathscr{F}\in\mathscr{E} \\ \|\mathscr{F}\|\leqslant 1}} |L_0(\mathscr{F})| = \sup_{f\in B^S} |l_0(f)| = \inf_{\lambda\in\mathscr{E}^{\perp}} \|L_0 - \lambda\|$$

$$= \inf_{\substack{m\in E^{\perp} \\ \Lambda\in X_1^*}} \left[\|l_0 - \tilde{l} - m\| + \varepsilon\|\Lambda\|\right].$$

The proof is complete.

REMARK 2.2. In Theorem 2.1 one can assume that Nf is a continuous operator from X to X_1 equipped with the weak topology. One can also assume that Nf acts only from E into X_1; then $\tilde{l}$ denotes functionals in X^* which coincide with

$\Lambda(Nf)$ on E. Instead of normed spaces one might consider linear locally convex topological spaces and assume that S_1 and S are convex bodies in those spaces. We will not write down the duality relations for those situations.

Let G be an n-connected finite domain bounded by rectifiable contours, and set $\Gamma = \partial G$. Let $\rho(\zeta) > 0$ be a given continuous function on Γ. We will consider the same classes of functions $C(\Gamma)$, $C_A(\Gamma)$, $C_\rho^1(\Gamma)$, $B_\rho^1(\Gamma)$ and $E_1(G)$ as in [55], §2.

THEOREM 2.3. *Let the following conditions be satisfied.*

1) $\omega(\zeta)$ *is a summable function on* Γ.

2) Nf *is a continuous linear operator defined on* $X = \tilde{\mathscr{M}}(\Gamma)$ *into a normed space* X_1, *and* S *is a sphere in* X_1 *of radius* ε *centered at the origin.*

3) B_ρ^S *and* C_ρ^S *are the sets of functions* f *in* B_ρ^1 *and* C_ρ^1 *respectively such that* $Nf \in S$.

4) *Every linear functional* $\tilde{l}(f) = \Lambda(N(f))$, $\forall \Lambda \in X_1^*$, *has the form*

$$\Lambda(Nf) = \tilde{l}(f) = \int_\Gamma f(\zeta)\,\omega_\Lambda(\zeta)\,d\zeta, \tag{2.11}$$

where $\omega_\Lambda(\zeta)$ *is a summable function on* Γ *defined by the functional* Λ.

Then

$$\sup_{f\in B_\rho^S}\left|\int_\Gamma f(\zeta)\,\omega(\zeta)\,d\zeta\right| = \sup_{f\in C_\rho^S}\left|\int_\Gamma f(\zeta)\,\omega(\zeta)\,d\zeta\right|$$
$$= \inf_{\substack{\Lambda\in X_1^*\\ \varphi(z)\in E_1(G)}}\left[\int_\Gamma \rho(\zeta)\,|\omega(\zeta) - \omega_\Lambda(\zeta) - \varphi(\zeta)|\,ds + \varepsilon\|\Lambda\|\right]. \tag{2.12}$$

Moreover, there exist $\Lambda^* \in X_1^*$ *and* $\varphi^*(z) \in E_1(G)$ *for which the infimum in* (2.12) *is attained.*

To prove the theorem, we must carry out the argument used in the proof of Theorem 2.2 in [55], using Theorem 2.1 of this section instead of the duality Theorem 1.1 in [55].

Now, let $\mathscr{D}$ be a subset of G, $\overline{\mathscr{D}} \subset G$. Let μ be a positive measure on $\mathscr{D}$. Let us take for the space X_1 in our problems the space $L^p(\mathscr{D}, \mu)$ consisting of functions $\psi(z)$ on $\mathscr{D}$ such that

$$\|\psi\|_{p,\mu} = \left\{\int_{\mathscr{D}} |\psi(z)|^p\,d\mu\right\}^{1/p} < +\infty. \tag{2.13}$$

Assume $p \geqslant 1$. For Nf we take the operator

$$Nf = \frac{1}{2\pi i}\int_\Gamma f(\zeta)\left(\sum_{k=0}^{r}\frac{c_k}{(\zeta - z)^{k+1}}\right)d\zeta, \tag{2.14}$$

where the c_k are given coefficients. The sphere S in $L_p(\mathscr{D}, \mu)$ consists of elements ψ such that

$$\|\psi\|_{p,\mu} \leqslant \varepsilon, \tag{2.15}$$

where ε is a given positive number.

In this case the class B_ρ^S consists of bounded analytic functions $f(z)$ in G such that

$$|f(\zeta)| \leqslant \rho(\zeta), \qquad \zeta \in \Gamma, \tag{2.16}$$

and

$$\int_{\mathscr{D}} \left| \sum_{k=0}^{r} \frac{c_k}{k!} f^{(k)}(z) \right|^p d\mu \leqslant \varepsilon. \tag{2.17}$$

It is clear that the hypothesis of Theorem 2.3 is satisfied. Then the duality relation (2.12) takes the following form ($p^{-1} + q^{-1} = 1$):

$$\sup_{f \in B_\rho^S} \left| \int_\Gamma f(\zeta)\omega(\zeta)\, d\zeta \right|$$

$$= \inf_{\substack{\Lambda \in L_q(\mathscr{D},\mu) \\ \varphi \in E_1(G)}} \left[\int_\Gamma \rho(\zeta) \cdot \left| \omega(\zeta) - \frac{1}{2\pi i} \int_{\mathscr{D}} \lambda(z) \left(\sum_{k=0}^{r} \frac{c_k}{(\zeta - z)^{k+1}} \right) d\mu - \varphi(\zeta) \right| ds \right.$$

$$\left. + \varepsilon \|\lambda\|_{q,\mu} \right]. \tag{2.18}$$

Indeed, an arbitrary linear functional over $L_p(\mathscr{D}, \mu)$ has the form

$$\Lambda(\psi) = \int_{\mathscr{D}} \psi\lambda\, d\mu, \qquad \lambda = \lambda(z) \in L_q(\mathscr{D}, \mu), \tag{2.19}$$

where

$$\|\Lambda\| = \|\lambda\|_{q,\mu}, \tag{2.20}$$

and finally

$$\tilde{l}(f) = \Lambda(Nf) = \int_{\mathscr{D}} \lambda(z) \left[\frac{1}{2\pi i} \int_\Gamma f(\zeta) \left(\sum_{k=0}^{r} \frac{c_k}{(\zeta - z)^{k+1}} \right) \right] d\mu$$

$$= \int_\Gamma f(\zeta) \left[\frac{1}{2\pi i} \int_{\mathscr{D}} \lambda(z) \left(\sum_{k=0}^{r} \frac{c_k}{(\zeta - z)^{k+1}} \right) d\mu \right] d\zeta.$$

(Using Fubini's theorem one can easily show that changing the order of integration is legitimate. Here it is essential that $\overline{\mathscr{D}} \subset G$.)

Therefore, we can set

$$\omega_\Lambda(\zeta) = \frac{1}{2\pi i} \int_{\mathscr{D}} \lambda(z) \left(\sum_{k=0}^{r} \frac{c_k}{(\zeta - z)^{k+1}} \right) d\mu \tag{2.21}$$

in (2.11). Instead of $L_p(\mathscr{D}, \mu)$ let us consider the space $C(\overline{\mathscr{D}})$ of continuous functions on $\overline{\mathscr{D}}$ (as above, $\overline{\mathscr{D}} \subset G$) with the metric

$$\|\psi\| = \max_{z \in \overline{\mathscr{D}}} \left| \frac{\psi(z)}{\rho_1(z)} \right|, \tag{2.22}$$

where $\rho_1(z) > 0$ is a given continuous weight function on $\overline{\mathscr{D}}$. Let Nf be defined by (2.14). Then the class B_ρ^S consists of bounded analytic functions in G satisfying (2.16) and such that

$$\left| \sum_{k=0}^{r} \frac{c_k}{k!} f^{(k)}(z) \right| \leqslant \rho_1(z). \tag{2.23}$$

(For S we take the unit sphere in $C(\overline{\mathscr{D}})$ with the metric (2.22).) The duality relation (2.12) takes the form

$$\sup_{f \in B_\rho^S} \left| \int_\Gamma f(\zeta)\omega(\zeta)\, d\zeta \right|$$

$$= \inf_{\substack{\varphi \in E_1(G) \\ \lambda}} \left[\int_\Gamma \rho(\zeta) \left| \omega(\zeta) - \frac{1}{2\pi i} \int_{\mathscr{D}} \sum_{k=0}^{r} \frac{c_k}{(\zeta - z)^{k+1}}\, d\lambda - \varphi(\zeta) \right| ds \right.$$

$$\left. + \int_{\mathscr{D}} \rho_1(z) |d\lambda| \right], \tag{2.24}$$

where $\lambda = \lambda(z)$ runs over the set of all complex Baire measures on $\mathscr{D}$. In that case the function $\omega_\Lambda(\zeta)$ in (2.11) has the form

$$\omega_\Lambda(\zeta) = \frac{1}{2\pi i} \int_{\overline{\mathscr{D}}} \sum_{k=0}^{r} \frac{c_k}{(\zeta - z)^{k+1}}\, d\lambda. \tag{2.25}$$

This follows immediately from the Cauchy formula and from changing the order of integration (see [26] or [48]).

Now, let us consider a problem where $\mathscr{D} \cap \Gamma \neq \varnothing$. Assume that $\mathscr{D}$ is closed in G and

$$\operatorname{meas}(\overline{\mathscr{D}} \cap \Gamma) = 0. \tag{2.26}$$

Define the norm (2.22) on $C(\overline{\mathscr{D}})$. Take the unit sphere in $D(\overline{\mathscr{D}})$ for S. Define the operator Nf from $C(\Gamma)$ into $C(\overline{\mathscr{D}})$ by

$$Nf = \psi(z) = \begin{cases} \dfrac{1}{2\pi} \displaystyle\int_\Gamma f(\zeta) \frac{\partial g(\zeta, z)}{\partial n}\, ds = \frac{1}{2\pi i} \int_\Gamma f(\zeta) p'_\zeta(\zeta, z)\, ds, & z \notin \Gamma \cap \overline{\mathscr{D}}, \\ f(z), & z \in \Gamma \cap \overline{\mathscr{D}}. \end{cases} \tag{2.27}$$

Here, $\partial g(\zeta, z)/\partial n$ is the normal derivative of the Green function of the domain G, and $p(\zeta, z)$ is the complex Green function of G. The corresponding class consists of continuous analytic functions in G satisfying (2.16) and such that

$$|f(z)| \leqslant \rho_1(z) \quad \text{on } \mathscr{D}. \tag{2.28}$$

We assume that

$$\rho_1(\zeta) \leqslant \rho(\zeta), \qquad \zeta \in \Gamma \cap \overline{\mathscr{D}}. \tag{2.29}$$

We cannot use Theorem 2.3 directly, because the representation (2.11) does not hold here. Instead we carry out the argument along the same scheme as in the proof of Theorem 2.3.

Consider a linear functional $\Lambda(\psi)$ on $C(\overline{\mathscr{D}})$. It has the form

$$\Lambda(\psi) = \int_{\overline{\mathscr{D}}} \psi(\varepsilon)\, d\lambda = \int_{\mathscr{D}} \psi(z)\, d\lambda + \int_{\overline{\mathscr{D}} \cap \Gamma} \psi(z)\, d\lambda = \Lambda_1(\psi) + \Lambda_2(\psi). \tag{2.30}$$

Here λ is a complex Baire measure on $\overline{\mathscr{D}}$, and

$$\|\Lambda\| = \int_{\overline{\mathscr{D}}} \rho_1(z)|d\lambda|. \tag{2.31}$$

We have

$$\begin{aligned} \tilde{l}(f) = \Lambda_1(Nf) &= \int_{\mathscr{D}} \left\{ \frac{1}{2\pi i} \int_{\Gamma} f(\zeta) p'_{\zeta}(\zeta, z)\, d\zeta \right\} d\lambda \\ &= \int_{\Gamma} f(\zeta) \left[\frac{1}{2\pi i} \int_{\mathscr{D}} p'_{\zeta}(\zeta, z)\, d\zeta \right] d\lambda. \end{aligned} \tag{2.32}$$

To justify the interchange of integrals we can refer to the variant of Fubini's theorem due to Tonelli (see [26], Chapter IV, §9.32, Theorem 2).

In fact, the function $(1/2\pi i) f(\zeta) p'_{\zeta}(\zeta, z)$ is obviously measurable on $\mathscr{D} \times \Gamma$ with respect to the measure $\lambda \times L$. (L is Lebesgue measure on Γ.) The integral

$$\int_{\mathscr{D}} \frac{1}{2\pi} \left[\int_{\Gamma} |f(\zeta)|\, |p'_{\zeta}(\zeta, z)|\, |d\zeta| \right] |d\lambda|$$

exists, since the function

$$u(z) = \frac{1}{2\pi} \int_{\Gamma} |f(\zeta) p'_{\zeta}(\zeta, z)|\, |d\zeta|$$

is bounded and harmonic. So the hypothesis of Tonelli's theorem is satisfied, and (2.32) has been justified. Here the function

$$\omega_{\Lambda}(\zeta) = \frac{1}{2\pi i} \int_{\mathscr{D}} p'_{\zeta}(\zeta, z)\, d\lambda \tag{2.33}$$

is summable on Γ. Moreover,

$$\tilde{l}_2(f) = \int_{\overline{\mathscr{D}} \cap \Gamma} \psi(\zeta)\, d\lambda = \int_{\overline{\mathscr{D}} \cap \Gamma} f(\zeta)\, d\lambda \tag{2.34}$$

(since $\psi(\zeta) = Nf = f(\zeta)$, $\zeta \in \Gamma \cap \overline{\mathscr{D}}$). Thus,

$$\tilde{l}(f) = \int_{\Gamma} f(\zeta) \omega_{\Lambda}(\zeta)\, d\zeta + \int_{\Gamma \cap \overline{\mathscr{D}}} f(\zeta)\, d\lambda. \tag{2.35}$$

Since a functional $m(f)$ vanishing on the subspace $E \subset C(\Gamma)$ consisting of functions analytic in G has the form

$$m(f) = \int_\Gamma f(\zeta)\varphi(\zeta)\,d\zeta, \qquad \varphi(\zeta) \in E_1(G),$$

by applying Theorem 2.1 we obtain

$$\begin{aligned}\sup_{f\in C_\rho^S} |l_0(f)| &= \sup_{f\in C_\rho^S}\left|\int_\Gamma f(\zeta)\omega(\zeta)\,d\zeta\right| \\ &= \inf_{\varphi\in E_1(G)} \left[\|l_0 - \tilde{l} - m\| + \|\Lambda\|\right],\end{aligned} \tag{2.36}$$

where

$$(l_0 - \tilde{l} - m)(f) = \int_\Gamma f(\zeta)[\omega(\zeta) - \omega_\Lambda(\zeta) - \varphi(\zeta)]\,d\zeta + \int_{\Gamma\cap\overline{\mathscr{D}}} f(\zeta)\,d\lambda. \tag{2.37}$$

Since (2.26), holds,

$$\|l_0 - \tilde{l} - m\| = \int_\Gamma \rho(\zeta)|\omega(\zeta) - \omega_\Lambda(\zeta) - \varphi(\zeta)|\,|d\zeta| + \int_{\Gamma\cap\overline{\mathscr{D}}} \rho_1(\zeta)|d\lambda|. \tag{2.38}$$

Using (2.38) and (2.31), we see that in taking the infimum in (2.36) we need to use only those functionals Λ for which the representing measure λ is concentrated in the interior of G (i.e., for which Λ vanishes on $\overline{\mathscr{D}} \cap \Gamma$). From (2.36) and (2.38) we obtain

$$\sup_{f\in C_\rho^S}\left|\int_\Gamma f(\zeta)\omega(\zeta)\,d\zeta\right| - \inf_{\varphi\in E_1(G)}\left[\int_\Gamma |\omega(\zeta) - \omega_\Lambda(\zeta) - \varphi(\zeta)|\,|d\zeta| + \int_{\mathscr{D}} \rho_1(z)\,|d\lambda|\right]. \tag{2.39}$$

Here λ runs over the set of all Baire measures concentrated on $\mathscr{D} \cap G$ and $\omega_\Lambda(\zeta)$ is defined by (2.33). Applying a standard argument, we show that here again

$$\sup_{f\in C_\rho^S}\left|\int_\Gamma f(\zeta)\omega(\zeta)\,d\zeta\right| = \sup_{f\in C_\rho^S}\left|\int_\Gamma f(\zeta)\omega(\zeta)\,d\zeta\right|. \tag{2.40}$$

We leave it to the reader to write out the duality relations for problems with additional conditions in spaces $X = E_p(G)$.

§3. Properties of extremal functions

From now on we assume that under the hypothesis of Theorem 2.3 the extremal function $f^*(z)$ does exist. In particular, this is the case for all concrete problems considered in §2, e.g., for problem (2.39), (2.40). The existence of the extremal functions in all these problems can easily be derived from compactness considerations. The following is now a characterization of extremals.

THEOREM 3.1. *For* $f^*(z) \in B_\rho^S$, $\Lambda^* \in X^*$ *and* $\varphi^*(z) \in E_1(G)$ *to be extremal for* (2.12), *it is necessary and sufficient that*

$$f^*(\zeta)[\omega(\zeta) - \omega_{\Lambda^*}(\zeta) - \varphi^*(\zeta)]\,d\zeta = e^{i\alpha}\rho(\zeta)|\omega(\zeta) - \omega_{\Lambda^*}(\zeta) - \varphi^*|\,ds \tag{3.1}$$

a.e. on Γ *and*

$$\Lambda^*(Nf^*) = \varepsilon e^{i\alpha}\|\Lambda^*\|; \tag{3.2}$$

α *in* (3.1) *and* (3.2) *is the same real number.*

Indeed,

$$\left|\int_\Gamma f(\zeta)\omega(\zeta)\,d\zeta\right| = \left|\int_\Gamma f^*(\zeta)[\omega(\zeta) - \omega_{\Lambda^*}(\zeta) - \varphi^*(\zeta)]\,d\zeta + \Lambda^*(Nf^*)\right|$$

$$\leqslant \int_\Gamma \rho(\zeta)|\omega(\zeta) - \omega_{\Lambda^*}(\zeta) - \varphi^*(\zeta)|\,ds + \varepsilon\|\Lambda^*\|. \tag{3.3}$$

On the other hand, because of (2.12), we have equalities in (3.3). From this (3.1) and (3.2) follow immediately.

In the case of problem (2.39), (2.40), which formally is not covered by Theorem 2.3, the equality (3.1) still holds for $f^*(z)$, the measure λ^* and $\varphi^*(z)$. Now we can render (3.2) concrete as follows:

$$f^*(z)\,d\lambda^* = \varepsilon e^{i\alpha}\rho_1(z)|d\lambda^*| \tag{3.4}$$

almost everywhere with respect to the measure $|d\lambda^*|$. The proof immediately follows from (2.39) and (2.40).

THEOREM 3.2. *If*

$$\omega(\zeta) - \omega_{\Lambda^*}(\zeta) - \varphi(\zeta) \neq 0 \quad \textit{almost everywhere on } \gamma, \tag{3.5}$$

where $\gamma \subset \Gamma$, $\operatorname{meas}\gamma > 0$, *then the extremal function* $f^*(z)$ *is determined uniquely up to a factor* $e^{i\beta}$.

PROOF. If $f^*(z)$ and $f_1^*(z)$ are two extremal functions, then (3.1) holds for each of them, where we can assume Λ^* and φ^* to be the same. (Only the α's for $f^*(z)$ and $f_1^*(z)$ may be different.) Therefore, $f^*(\zeta) = e^{i\beta}f_1^*(\zeta)$ on γ, and then, according to the Luzin-Privalov theorem, $f^*(z) = e^{i\beta}f_1^*(z)$.

(3.1) allows us to carry out a complete investigation of the properties of the extremal function $f^*(z)$ by the same means as in [55]. The additional function $\omega_{\Lambda^*}(\zeta)$ in (3.1) does not complicate matters too greatly, since it is analytic on $\overline{\mathscr{D}}$ in the most important concrete problems (see (2.21), (2.25) and (2.33)). Let us state some of the results (cf. §§3–7 in [55]).

THEOREM 3.3. *Let* $\gamma \subset \Gamma$ *be an analytic arc, and a.e. on* γ *let* $\omega(\zeta)$ *and* $\omega_{\Lambda^*}(\zeta)$ *coincide with boundary values of analytic functions* $\omega(\zeta)$ *and* $\omega_{\Lambda^*}(\zeta)$ *which belong to the class* $E_1(\tilde G)$ *in a domain* $\tilde G$ *attached to* γ. *Let* (3.5) *hold. Then the product* $f^*(z)[\omega(z) - \omega_{\Lambda^*}(z) - \varphi(z)]$ *can be analytically continued across* γ.

THEOREM 3.4. *Let all the hypotheses of Theorem 3.3, except possibly the analyticity of contours, be satisfied for the whole boundary* Γ. *Then* $f^*(z)$ *has the form*

$$f^*(z) = e^{i\alpha} \exp\left[\frac{1}{2\pi}\int_\Gamma \ln\rho(\zeta)\frac{\partial p(z,\zeta)}{\partial n}\,ds\right]\exp\left[-\sum_{k=1}^{m} p(z,z_k)\right], \quad (3.6)$$

where $p(z,\zeta)$ *is the complex Green function of* G *and* $z_1,\dots,z_m$ *are zeros of* $f^*(z)$.

In particular, we obtain the following theorem.

THEOREM 3.5. *If* $\rho(\zeta)\equiv 1$ *under the assumptions of Theorem 3.4, then either the extremal function* $f^*(z)$ *gives a mapping of* G *onto the* $m \geqslant n$ *times covered unit disk or* $f^*(z) = e^{i\alpha}$.

Thus, the mapping property of the extremal function $f^*(z)$ established for problems without additional constraints also holds for many problems with additional constraints. Roughly speaking, if $\mathcal{D}$, the set where these constraints are defined, does not come out on the boundary, then the mapping property of the extremal function in the class B^1 is preserved.

In order to satisfy the hypotheses of the theorems in this section, one can take, for instance,

$$\omega(\zeta) = \sum_{k=1}^{r} \frac{\beta_k}{(\zeta-\alpha_k)^{n_k}}, \quad (3.7)$$

where $\alpha_1,\dots,\alpha_r$ are arbitrary points in G which are not separated from Γ by any component of $\mathcal{D}$. (Here the β_k are given coefficients, and the $n_k \geqslant 1$ are given integers.) If we do not require the last condition for the function (3.7), then (3.5) may fail and the above theorems are no longer true.

For example, let us take for G the unit disk $|z| < 1$ and $\rho(\zeta) \equiv 1$. Consider the class B_ρ^S consisting of functions $f(z)$ satisfying the following inequalities:

$$|f(z)| \leqslant 1, \quad |z| < 1,$$
$$|f(z)| \leqslant \varepsilon, \quad |z| = r < 1.$$

Let us consider the problem of $\sup|2\pi f(0)|$ in this class. For this problem, clearly $\omega(\zeta) = 1/\zeta$. From the maximum principle we obtain that the function $f^*(z) \equiv \varepsilon$ is extremal, which contradicts the theorems in this section. This occurs precisely because condition (3.5) is not satisfied. Indeed, for $|\zeta| > 0$ (in particular, on the circle $|\zeta| = 1$), the following identity holds:

$$\frac{1}{\zeta} = \frac{1}{2\pi i}\int_{|z|=r} \frac{z^{-1}\,dz}{z-\zeta}. \quad (3.8)$$

Therefore, $1/\zeta$ coincides with one of the functions $\omega_\Lambda(\zeta)$ (see (2.25)). From (2.24) we have for our problem that

$$2\pi \sup|f(0)| = \inf_{\substack{\varphi\in E_1\\ \lambda}} \left[\int_\Gamma \left|\frac{1}{\zeta} - \frac{1}{2\pi i}\int_{|z|=r}\frac{d\lambda}{z-\zeta} - \varphi(\zeta)\right| ds + \varepsilon\cdot\int_{|z|=r}|d\lambda|\right], \quad (3.9)$$

where $\lambda(l)$ is an arbitrary Baire measure on the circle $|z| = r$. The function $\lambda^*(l)$ defined by $d\lambda^* = z^{-1}\,dz$ (i.e., taken from (3.8)) is extremal in (3.9) (with $\varphi^*(z) \equiv 0$). This is seen from the fact that for this function the first integral in (3.9) vanishes (by (3.8)) and the second is equal to

$$\int_{|z|=r} |d\lambda| = 2\pi.$$

Thus, the extremal value of the right-hand side is equal to $2\pi\varepsilon$, in complete agreement with the maximum principle.

Theorem 3.5 gives an estimate from below of the number m of factors $\exp[-p(z, z_k)]$ in the representation (3.6) of the extremal function $f^*(z)$ (for $\rho(\zeta) \equiv 1$).

In many important particular cases, it is also possible to estimate m from above.

THEOREM 3.6. *Let $\alpha_1, \ldots, \alpha_r$ be distinct points in G. Let B_ρ^S consist of functions which belong to B_ρ and satisfy either the additional constraints*

$$\left|\sum_{j=0}^{k} \frac{c_j}{j!} f^{(j)}(\alpha_\nu)\right| \leqslant \varepsilon_\nu \qquad (\nu = 1, \ldots, r), \tag{3.10}$$

where arbitrary complex numbers c_j $(j = 0, \ldots, k)$ and numbers $\varepsilon_\nu \geqslant 0$ $(\nu = 1, \ldots, r)$ are given, or the additional constraint

$$\sum_{\nu=1}^{r} \mu_\nu \left|\sum_{j=0}^{k} \frac{c_j}{j!} f^{(j)}(\alpha_\nu)\right|^p \leqslant \varepsilon \tag{3.11}$$

with prescribed c_j $(j = 0, \ldots, k)$, $\mu_\nu \geqslant 0$ $(\nu = 1, \ldots, r)$, $p \geqslant 1$ and $\varepsilon > 0$. Let there also be given points $\beta_1, \ldots, \beta_k$ different from $\alpha_1, \ldots, \alpha_r$, nonnegative integers n_i $(i = 1, \ldots, R)$, and complex numbers $\gamma_{i\nu}$ $(i = 1, \ldots, R;\ \nu = 0, \ldots, n_i)$. Then the extremal function in the problem of

$$\sup_{f \in B_\rho^S} \left|\sum_{i=1}^{R} \sum_{\nu=0}^{n_i} \gamma_{i\nu} f^{(\nu)}(\beta_i)\right| \tag{3.12}$$

has the form (3.6) *with m satisfying the inequality*

$$m \leqslant \sum_{i=1}^{R} (n_i + 1) + (k + 1)r + n - 2. \tag{3.13}$$

In particular, if $\rho(\zeta) \equiv 1$, then $f^*(z)$ maps G onto the m-sheeted unit disk and

$$n \leqslant m \leqslant \sum_{i=1}^{R} (n_i + 1) + (k + 1)r + n - 2, \tag{3.14}$$

whenever $f^*(z) \not\equiv e^{i\alpha}$.

PROOF. The additional constraints (3.10) or (3.11) used in the definition of the class B_ρ^S are particular cases of (2.23) and (2.17) respectively. The function $\omega(\zeta)$

in (3.12) has the form

$$\omega(\zeta) = \sum_{i=1}^{R} \sum_{\nu=0}^{n_i} \frac{\gamma_{i\nu}}{(\zeta - \beta_i)^{\nu+1}}.$$

Applying the argument principle to the positive differentials in the left-hand sides of (3.1), we obtain (3.13). The estimate (3.13), together with Theorem 3.5, implies (3.14).

§4. Further properties of extremal functions. Tchebycheff's sets

So-called refinement theorems (see, for instance, [18], Chapters 7 and 9) play an important role in the theory of uniform approximation. It turns out that analogues of this fact can also be observed for some extremal problems in the theory of analytic functions.

In this section G is the unit disk $|z| < 1$. Let $\mathscr{D}$ be a closed subset in G and let $\rho_1(z)$ be a continuous function on $\mathscr{D}$. For the class B_ρ^S we take the class of functions satisfying the inequalities

$$|f(\zeta)| \leqslant \rho(\zeta), \quad |\zeta| = 1, \tag{4.1}$$

$$|f(z)| \leqslant \rho_1(z), \quad z \in \mathscr{D}. \tag{4.2}$$

To underline the dependence of this class on the set $\mathscr{D}$ we will denote it by $B_\rho^{S(\mathscr{D})}$.

THEOREM 4.1. *Assume that the following conditions hold*:

1) $\mathscr{D} \cap \Gamma = \varnothing$.

2) $\rho_1(z)$ *is positive, continuous on* $\overline{\mathscr{D}}$ *and harmonic at interior points of* $\mathscr{D}$.

3) *The function* $\omega(\zeta)$ *is the boundary value on* Γ ($|\zeta| = 1$) *of a function* $\omega(z)$ *analytic in an annulus* $\tilde{G}$ *attached to* Γ *and belonging to* $E_1(\tilde{G})$ *there.*

4) *For any* $\varphi(z) \in H_1(G) = E_1(G)$ *and any Baire measure* $\lambda(e)$ *defined on the boundary* γ *of* $\mathscr{D}$,

$$\omega(\zeta) - \int_{\mathscr{D}} \frac{d\lambda}{z - \zeta} - \varphi(\zeta) \neq 0 \quad \text{a.e. on } \Gamma. \tag{4.3}$$

5) *For any* m *and arbitrary* $\alpha_1, \dots, \alpha_m$, $|\alpha_j| < 1$ $(j = 1, \dots, m)$, *the equation*

$$\prod_{j=1}^{m} \left| \frac{z - \alpha_j}{1 - \bar{\alpha}_j z} \right| \exp \frac{1}{2\pi} \int_0^{2\pi} \ln \rho(e^{i\theta}) \frac{1 - r^2}{1 + r^2 - 2r\cos(\theta - \varphi)}\, d\theta = \rho_1(z) \tag{4.4}$$

$(z = re^{i\varphi})$ *has only finitely many solutions on* γ.

Then there exists a finite set $\mathscr{D}_N \subset \gamma$ *such that the solutions of the problem of*

$$\sup \left| \int_\Gamma f(\zeta)\omega(\zeta)\, d\zeta \right| \tag{4.5}$$

in the classes $B_\rho^{S(\mathscr{D})}$ *and* $B_\rho^{S(\mathscr{D}_N)}$ *coincide.*

PROOF. Since $\rho_1(z)$ is assumed to be harmonic in the interior of $\mathscr{D}$, the classes $B_\rho^{S(\mathscr{D})}$ and $B_\rho^{S(\gamma)}$ coincide. The characterization of the extremal function $f^*(z)$

(see (3.1) and (3.4)) in the class $B_\rho^{S(\gamma)}$ in our case has the form

$$\begin{aligned} f^*(\zeta)\left[\omega(\zeta) - \frac{1}{2\pi i}\int_\gamma \frac{d\lambda}{z-\zeta} - \varphi^*(\zeta)\right] d\zeta \\ = e^{i\alpha}\rho(\zeta)\left|\omega(\zeta) - \frac{1}{2\pi i}\int_\gamma \frac{d\lambda}{z-\zeta} - \varphi^*(\zeta)\right| ds, \end{aligned} \tag{4.6}$$

$$f^*(z)\, d\lambda^* = e^{i\alpha}\rho_1(z)|d\lambda^*| \tag{4.7}$$

almost everywhere with respect to the measure $|d\lambda^*|$ on γ. In particular, it follows from (4.7) that

$$|f^*(z)| = \rho_1(z) \tag{4.8}$$

almost everywhere with respect to $|d\lambda^*|$ on $\mathscr{D}$. Since all the hypotheses of Theorem 3.4 hold, $f^*(z)$ has the form (3.6), and therefore, (4.8) has the form (4.4). Hence, as assumed, (4.8) has finitely many roots $\alpha_1, \dots, \alpha_N$ on γ. Therefore the measure $d\lambda^*$ is concentrated only at these points and hence we can replace integration over γ in (4.6) by integration over $\mathscr{D}_N = (\alpha_1, \dots, \alpha_N)$. This means that (4.6) and (4.7) also characterize the extremal for (4.5) in the class $B_\rho^{S(\mathscr{D}_N)}$. The theorem is proved.

It is natural to call $\mathscr{D}_N$ the *Tchebycheff set* of our problem. Let us point out an important special case of the above theorem.

THEOREM 4.2. *Under the hypothesis of the previous theorem, assume that $\rho(\zeta) \equiv 1$, and γ consists of finitely many disjoint circular arcs and line segments. Let $\rho_1(z)$ be constant on every circle and every segment. Also, if γ_j is an arc with center located in the interior of G and noneuclidean radius $r_j < 1$*[2] *and ε_j is the value of $\rho_1(z)$ on γ_j, assume that for any integer $p \geqslant 1$*

$$r_j^p \neq \varepsilon_j. \tag{4.9}$$

Then the conclusion of Theorem 4.1 holds. Moreover, if N is the number of points in $\mathscr{D}_N$, m is the number of sheets of the multisheeted disk onto which $f^*(z)$ maps G, l is the number of full circles included in γ and l_1 is the number of arcs in γ that are not full circles, then

$$N \leqslant lm + (m+1)l_1. \tag{4.10}$$

If, in addition,

$$\omega(\zeta) = \sum_{i=1}^{R}\sum_{\nu=1}^{n_i} \frac{\gamma_{i\nu}}{(\zeta - \beta_i)^\nu}, \tag{4.11}$$

then

$$N \geqslant m + 1 - \sum_{i=1}^{R} n_i. \tag{4.12}$$

---

(2) In contrast to generally accepted definitions, we mean by r_j the radius of the circle $|w| = r_j$, the image of the circle $c_j \supset \gamma_j$ under the mapping $w = (z-\alpha)/(1-\bar{\alpha}z)$, where α is the noneuclidean center of c_j.

PROOF. According to Theorem 3.4, where now $\rho(\zeta) \equiv 1$, the extremal function $f^*(z)$ has the form

$$f^*(z) = \prod_{\nu=1}^{m} \frac{z - \alpha_\nu}{1 - \bar{\alpha}_\nu z}, \qquad |\alpha_\nu| < 1. \tag{4.13}$$

First let us consider the case where γ_j is one of those arcs in γ with noneuclidean center β in G and radius $r_j < 1$. Assume that the point β occurs among the points α_ν in (4.13) m_1 times (possibly $m_1 = 0$). Making the change of variable

$$t = \frac{z - \beta}{1 - \bar{\beta} z}, \tag{4.14}$$

we transform $f^*(z)$ into the function

$$F(t) = t^{m_1} \prod_{\nu=1}^{m-m_1} \frac{t - a_\nu}{1 - \bar{a}_\nu t}, \tag{4.15}$$

where $a_1, \ldots, a_{m-m_1}$ are images of $\alpha_\nu \neq \beta$ under the transformation (4.14). The arc γ_j is transformed into the arc Γ_j centered at $t = 0$ with (euclidean) radius r_j. By (4.8)

$$r_j^{m_1} \prod_{\nu=1}^{m-m_1} \left| \frac{t - a_\nu}{1 - \bar{a}_\nu t} \right| = \varepsilon_j < 1$$

or

$$r_j^{2m_1} \prod_{\nu=1}^{m-m_1} \left| r_j e^{i\theta} - a_\nu \right|^2 - \varepsilon_j^2 \prod_{\nu=1}^{m-m_1} \left| 1 - \bar{a}_\nu e^{i\theta} r_j \right|^2 = 0 \tag{4.16}$$

at all the images of the points of increase of the measure $d\lambda^*$ (see (4.7)) lying on Γ_j. Because of (4.9) this equation cannot be reduced to $r_j^{m_1} = \varepsilon_j$, hence its left-hand side is a trigonometric polynomial, not identically zero, of degree $m - m_1 \leqslant m$. Since, according to the definition of $B_1^{S(\mathscr{D})}$, $|f^*(z)| \leqslant \varepsilon_j$ at all points of γ_j, the polynomial in the left-hand side of (4.16) is nonpositive on Γ_j. Therefore, (4.16) has at most m distinct roots if γ_j is a full circle, and at most $m + 1$ roots if γ_j is an arc.

In all cases different from those considered above, none of the factors constituting the Blaschke product (4.13) can have a constant modulus on γ_j. Therefore, (4.8) now gives us

$$\prod_{\nu=1}^{m} |z - \alpha_\nu|^2 - \varepsilon_j^2 \prod_{\nu=1}^{m} |1 - \bar{\alpha}_\nu z|^2 = 0, \qquad \varepsilon_j < 1, \tag{4.17}$$

and the left-hand side of (4.17) is a nonpositive trigonometric polynomial([3]) on γ_j, not identically zero. (Assuming that the polynomial on the left-hand side of (4.17) is identically zero on γ_j, we would conclude that it is identically zero on the whole circle c_j containing γ_j. But this circle must have common points with the

([3])If β is the center of γ_j and r is the radius, then for $z \in \gamma_j$ we have $z = \beta + re^{i\theta}$, $0 \leqslant \theta \leqslant 2\pi$. This implies that (4.17) is a trigonometric polynomial θ.

circle $|z| = 1$. This is a contradiction, since according to our hypothesis $|f^*(z)| \leqslant \varepsilon_j < 1$ on c_j and $|f^*(z)| = 1$ on $|z| = 1$.) It follows again that the number of distinct roots of (4.17) does not exceed $m + 1$. Finally, if γ_j is a segment of the straight line $y = kx + b$, then on the left-hand side of (4.17) we have a nonnegative (algebraic) polynomial on γ_j of degree $2m$, which has at most $m + 1$ distinct roots on γ_j. (The same argument as above shows that the polynomial on the left-hand side of (4.17) is not identically zero.) All the above reasonings imply (4.10). (4.12) follows from (3.13).

COROLLARY 4.3. *Assume that under the hypothesis of the previous theorem* γ *consists of one arc or one line-segment and*

$$\omega(\zeta) = 1/2\pi i(\zeta - \beta) \quad \left(\textit{problem of } \sup|f(\beta)|\right). \tag{4.18}$$

Then the number N *of points in the Tchebycheff set and the number* m *of sheets of the multisheeted disk* $|w| < 1$ *onto which the extremal function* $w = f^*(z)$ *maps the disk are related by the inequalities*

$$m \leqslant N \leqslant m + 1. \tag{4.19}$$

If γ *is a full circle, then*

$$m = N. \tag{4.20}$$

Indeed, from (4.10) we obtain $N \leqslant m + 1$. From (4.12) it follows that $N \geqslant m$. Thus, $m \leqslant N \leqslant m + 1$, or, if γ is a full circle, $m = N$. Q.E.D.

We complement the results of §3 with a theorem concerning the form of extremal functions in the class defined by inequalities (4.1) and (4.2) provided that $\overline{\mathscr{D}} \cap \Gamma \neq \varnothing$. We put this theorem here because it deals with the problem of the distribution of points where (4.7) holds.

First let us note that if

$$\rho(re^{i\theta}) < \exp\left[\frac{1}{2\pi}\int_0^{2\pi} \ln \rho(\theta) \frac{1 - r^2}{1 + r^2 - 2r\cos(\theta - \varphi)}\, d\theta\right], \tag{4.21}$$

then (4.2) gives an additional restriction in comparison to (4.1).

THEOREM 4.4. *Let* meas $\overline{\mathscr{D}} \cap \Gamma = 0$, *and assume that the part of* $\mathscr{D}$ *contained in a neighborhood of any of the points* $\zeta \in \overline{\mathscr{D}} \cap \Gamma$ *lies inside an angle formed by chords issuing from this point. Assume also that* $\rho_1(z)$ *is positive and is continuous on* $\mathscr{D}$, $\rho(e^{i\theta})$ *is continuous on* Γ *and* (4.2) *holds in the stronger form*

$$\rho_1(re^{i\varphi}) < \exp\left[\frac{1}{2\pi}\int_0^{2\pi} \ln \rho(\theta) \frac{1 - r^2}{1 + r^2 - 2r\cos(\theta - \varphi)}\, d\theta\right] - \delta, \tag{4.22}$$

where $\delta > 0$ *does not depend on points* $re^{i\varphi} \in \mathscr{D}$. *Then the extremal function* $f^*(z)$ *in the problem of* $\sup|\int_\Gamma f(\zeta)\omega(\zeta)\, d\zeta|$ *with* $\omega(\zeta)$ *satisfying the hypotheses of Theorems* 4.1 *and* 4.2 *has the following form*:

$$f^*(z) = e^{i\alpha} \prod_{\nu=1}^{\infty} \frac{z - \alpha_\nu}{1 - \bar{\alpha}_\nu z} \frac{\alpha_\nu}{|\alpha_\nu|} \cdot \exp \frac{1}{2\pi} \int_0^{2\pi} \frac{e^{i\theta} + z}{e^{i\theta} - z} \ln \rho(\theta)\, d\theta. \tag{4.23}$$

Moreover, each of the points $\zeta \in \overline{\mathscr{D}} \cap \Gamma$ *is a cluster point of the zeros of the Blaschke product in* (4.23), *and there are no other cluster points. Each of the points of* $\overline{\mathscr{D}} \cap \Gamma$ *is also a cluster point for the points of increase of the extremal measure* $d\lambda^*$ (*i.e., the points where* (4.7) *holds*).

PROOF. From (2.39) and (2.40) it follows that extremal functions $f^*(z)$ and $\varphi^*(z)$ and the measure λ^* are related by

$$\begin{aligned} f^*(\zeta)\left[\omega(\zeta) - \frac{1}{2\pi i}\int_{\mathscr{D}} p'_\zeta(\zeta, z)\,d\lambda^* - \varphi^*(\zeta)\right] d\zeta \\ = e^{i\alpha}\rho(\zeta)\left|\omega(\zeta) + \frac{1}{2\pi i}\int_{\mathscr{D}} p'_\zeta(\zeta, z)\,d\lambda^* - \varphi^*(\zeta)\right| d\theta. \end{aligned} \tag{4.24}$$

(This relation is of the same type as (3.1).) As to (4.7), it preserves its form. Let us now show that each of the points of $\overline{\mathscr{D}} \cap \Gamma$ is a cluster point for the points of growth of the measure $d\lambda^*$. Suppose this is not true for a point ζ_0. Then the function

$$\omega_{\Lambda^*}(\zeta) = \frac{1}{2\pi i}\int_{\mathscr{D}} p'_\zeta(\zeta, z)\,d\lambda^* \tag{4.25}$$

is analytic on an arc C of a circle containing the point ζ_0 inside. Then, in some domain $\tilde{\tilde{G}} \subset G$ attached to Γ,

$$|f^*(z)| = F_1(z)\exp\left[\frac{1}{2\pi}\int_0^{2\pi} \ln\rho(e^{i\theta})\frac{1 - r^2}{1 + r^2 - 2r\cos(\theta - \varphi)}\,d\theta\right], \tag{4.26}$$

where $F_1(z)$ is continuous in the closure of $\tilde{\tilde{G}}$ and $|F_1(\zeta)| = 1$ on Γ. This immediately implies that at points of $\mathscr{D}$ sufficiently close to ζ_0, the following inequality holds:

$$|f^*(z)| > \exp\left[\frac{1}{2\pi}\int_0^{2\pi} \ln\rho(e^{i\theta})\frac{1 - r^2}{1 + r^2 - 2r\cos(\theta - \varphi)}\,d\theta\right] - \frac{\delta}{2}. \tag{4.27}$$

But since $|f^*(z)| \leqslant \rho_1(z)$ and (4.22) holds, we arrive at the desired contradiction.

From the analyticity of $\omega_\Lambda(\zeta)$, in (4.25), on arcs complementary to $\overline{\mathscr{D}} \cap \Gamma$, it follows that $|f^*(\zeta)| = \rho(\zeta)$ almost everywhere on Γ. The Smirnov parametric representation gives us

$$f^*(z) = e^{i\alpha}\prod_{\nu=1}^{\infty}\frac{z - \alpha_\nu}{1 - \bar{\alpha}_\nu z}\cdot\frac{\alpha_\nu}{|\alpha_\nu|}\exp\left\{\frac{1}{2\pi}\int_0^{2\pi}\frac{e^{i\theta} + z}{e^{i\theta} - z}\left[\ln\rho(e^{i\theta})\,d\theta + d\mu\right]\right\}, \tag{4.28}$$

where $d\mu \leqslant 0$ is a singular measure.

Applying (4.26) to the arc $L \subset \Gamma \setminus \overline{\mathscr{D}} \cap \Gamma$, we see that in (4.26) $F_1(z)$ is a product of the two last factors in (4.28). From the properties of the function $F_1(z)$ it follows, first of all, that the Blaschke product cannot have any cluster points of the set of its zeros on L. Furthermore, the harmonic function

$$u(re^{i\varphi}) = \frac{1}{2\pi}\int_0^{2\pi} \frac{1-r^2}{1+r^2-2r\cos(\theta-\varphi)}\,d\mu \tag{4.29}$$

is continuous in the closure of the domain $\tilde{\tilde{G}}$ attached to L and vanishes on L. From the way in which the measure $d\mu$ is recovered in terms of values of the function $u(re^{i\theta})$, we conclude that the measure $d\mu$ in (4.29) can be chosen in such a way that it does not have any points of increase on L. Thus, the measure $d\mu$ in (4.28) is concentrated on $\overline{\mathscr{D}} \cap \Gamma$.

If $d\mu \not\equiv 0$, then according to a theorem of de la Vallée-Poussin on the derivatives of functions of bounded variation (see [48], Chapter IV, §9, Theorem 9.6), there exists a point $\zeta_0 \in \overline{\mathscr{D}} \cap \Gamma$ such that $\mu'(\zeta_0) = -\infty$. But then from (4.28) it follows that

$$f^*(z) \to 0 \quad \text{as } z \to \zeta_0, \qquad z \in \mathscr{D}. \tag{4.30}$$

This is impossible, since (4.7) must hold at the points of increase of the measure λ^*, and in view of continuity of $\rho_1(z)$ on $\overline{\mathscr{D}}$,

$$\rho_1(z) \geqslant m > 0. \tag{4.31}$$

But we have already shown that the points of increase of $|d\lambda^*|$ accumulate at ζ_0.

Formula (4.23) is established. It remains to show that each of the points $\zeta \in \overline{\mathscr{D}} \cap \Gamma$ is a cluster point for the zeros of the Blaschke product in (4.23). Assuming this is false, we would obtain that

$$\left| \prod_{\nu=1}^{\infty} \frac{z-\alpha_\nu}{1-\bar{\alpha}_\nu z}\,\frac{\alpha_\nu}{|\alpha_\nu|} \right| \to 1, \qquad z \to \zeta,$$

which again contradicts (4.22). The proof is complete.

NOTES

Chapter I

§1. The Carathéodory-Fejér problem was considered by those authors in [27]. Their main result is Theorem 2.3 (see below). A slightly more general interpolation problem in the disk was studied by Pick [45], [46] and R. Nevanlinna [40], [41]. In §1 we follow [51]. In this connection, let us point out Garkavi's paper [11], where he studied the dual of the Carathéodory-Fejér problem (he calls it the Helly problem). Theorem 1.8 is contained in Grunsky's paper [20].

§2. Theorems 2.1 and 2.2 are related to §1 and §§2–4 of [55]. For $\rho(\theta) \equiv 1$, $p = 1$ and the functionals (2.1) with $\beta = 0$, the problem in Theorem 2.2 is the Carathéodory-Fejér problem. For the functionals (2.2) it is the Nevanlinna-Pick problem. The main Theorem 2.3 is due to Carathéodory and Fejér in the case of Taylor coefficients and to Pick in the case of the functionals (2.2). Our presentation is based on [17] (see also [15] and [16]). Other presentations and generalizations can be found in the monographs [19] and [4] and the paper [3]. Far-reaching generalizations of the algebraic aspects of the Carathéodory-Fejér problem have been given by Adamyan, Arov and Kreĭn in [31] and [32].

§3. Theorems 3.1 and 3.2 are due to Schur [61]. Theorem 3.3 is from Akhiezer and Kreĭn [6]. The algorithm presented in Theorem 3.4 is also due to Schur [61]. Theorems 3.5, 3.7 and 3.8 are presented in [44]. Theorem 3.9 is taken from Goluzin's paper [16]. Here we have just slightly touched on some facts of the extensive theory of coefficients in various function classes, having in mind to give further examples of the use of the duality method. Further details on coefficients can be found in [7], [23] and [47].

§4. Theorem 4.1 was proved in [2]. Some related problems on radii of convexity and starlikeness in the class $B_{\rho(\zeta)}$ and related classes (in multiply connected domains) were considered in [43] and [52]. The calculation of the radius of univalence in case of the unit disk is due to Landau [36] (see also [8]). Dieudonné [22] and Cacridis-Theodorakopulos [25] solved the problem of the precise value of the radii of p-valence, starlikeness and convexity in the class H_∞ in the disk. The calculations and estimates of the corresponding radii for various subclasses in H_p, $p \geqslant 1$, can be found in [34] and [35].

Chapter II

§1. Results 1.1–1.5 were proved by Kabaĭla [24]. The form of extremal function in problem (1.13) had been found earlier by Khavinson in an unpublished paper (this is mentioned in [50]).

§2. The results of this section are due to Khavinson. In [49] this method was extended to more general Orlicz type classes. Various estimates for the classes H_p and H_p^0, $0 < p < 1$, were studied in [14], [34] and [35]. The spaces H_p, $0 < p < 1$, cannot be furnished with a norm, as was shown in [37]. However, they turned out to be closely related to certain normed spaces (see [21]). In particular, this connection allowed us to establish the form of the linear functionals on H_p.

We have not touched on questions related to the class $\mathscr{P}$ of analytic functions with positive real part, although the extrema in that class are closely related to the problems studied here. Representing those functions by the Schwarz integral, one can reduce extremal problems in $\mathscr{P}$ to corresponding extremal problems in the theory of moments. These questions are studied in the monographs [5] and [33] in great detail. A survey of other methods applied here can be found in [1]. For multiply connected domains, extremal problems in $\mathscr{P}$ have been considered by Heins [58], Nehari [42] and Kochetkov [29]. Structural formulas similar to the Smirnov parametric representation connect the class $\mathscr{P}$ with other interesting classes of functions. In particular, this connection was used to study extremal problems in the class A in the disk in [12], [30] and [13].

Chapter III

§1. There is a vast literature treating problems of estimating analytic functions satisfying a specific restriction on growth in the domain and additional growth constraints on a certain subset. A detailed survey of those investigations is given by the author in [54]. In all those papers the major concern is how to obtain the best possible estimate on the growth of functions of the class under study. The nature and properties of extremal functions remained unexplored.

In [56] and [57] Heins demonstrated another approach to Milloux-type problems. He concentrated on studying the properties of extremal functions and, if possible, on exact determination of this extremal function, and, therefore, the corresponding extremum . Heins's work was continued in [38], where an analogue of the Milloux problem in the annulus was considered. In Khavinson's papers [53] and [54] the duality method has been extended to these problems (in the class of bounded functions or classes E_p with additional conditions). The detailed account of this method (under slightly more general assumptions than in the present paper) is contained in the survey [54]. Further generalizations are due to Kochetkov [28] and Hejhal [59].

Milloux's theorem was proved in [39]. In [60] Schmidt has found that $1/\pi$ is the best constant. The reduction used in the proof of Theorem 1.1 is due to Heins [56].

§§2–4. The results are due to Khavinson [53] and [54].

REFERENCES

1. Yu. E. Alenitsyn, G. V, Kuz'mina and N. A. Lebedev, *Methods and results of geometric function theory*, Supplement to [15], pp. 532–626 (English pp. 563–672).

2. Yu. E. Alenitsyn and S. Ya. Khavinson, *The radius of p-valence for bounded analytic functions in multiply-connected domains*, Mat. Sb. **52(94)** (1960),653–657. (Russian)

3. N. I. Akhiezer, *On a minimum-problem in function theory, and on the number of roots of an algebraic equation which lie in the unit disk*, Izv. Akad. Nauk SSSR Ser. Mat. Estestv. Nauk **1931**, 1169–1189. (Russian)

4. ______, *Lectures on the theory of approximation*, 2nd rev. ed., "Nauka", Moscow, 1965; English transl. of 1st ed., Ungar, New York, 1956.

5. ______, *The classical moment problem and some related questions in analysis*, Fizmatgiz, Moscow, 1961; English transl., Oliver & Boyd, London, and Hafner, New York, 1965.

6. N. I. Akhiezer and M. G. Kreĭn, *Some questions in the theory of moments*, Gos. Nauchno-Tekhn. Izdat. Ukraïn., Kharkov, 1937; English transl., Amer. Math. Soc., Providence, R. I., 1962.

7. N. K. Bari, *Trigonometric series*, Fizmatgiz, Moscow, 1961; English transl., Vols. I, II, Pergamon Press, Oxford, and Macmillan, New York, 1964.

8. Georges Valiron, *Fonctions analytiques*, Presses Univ. France, Paris, 1954.

9. F. R. Gantmakher, *The theory of matrices*, GITTL, Moscow, 1953; English transl., Vols. 1, 2, Chelsea, New York, 1959.

10. F. R. Gantmakher and M. G. Kreĭn, *Oscillation matrices and kernels and small oscillations of mechanical systems*, 2nd ed., GITTL, Moscow, 1950; English transl., U. S. Atomic Energy Commission, Oak Ridge, Tenn., 1961; German transl., Akademie-Verlag, Berlin, 1960.

11. A. L. Garkavi, *The Helly problem and best approximation in a space of continuous functions*, Izv. Akad. Nauk SSSR Ser. Mat. **31** (1967), 641–656; English transl. in Math. USSR Izv. **1** (1967).

12. S. A. Gel'fer, *A variational method in the theory of functions of bounded characteristic*, Mat. Sb. **69(111)** (1966), 422–433; English transl. in Amer. Math. Soc. Transl. (2) **72** (1968).

13. V. P. Vazhdaev and S. A. Gel'fer, *Some estimates in the class of analytic functions of bounded type*, Mat. Sb. **84(126)** (1971), 273–289; English transl. in Math. USSR Sb. **13** (1971).

14. S. A. Gel'fer and L. V. Kresnyakova, *A variational method in the theory of analytic functions with bounded mean modulus*, Mat. Sb. **67(109)** (1965), 570–585. (Russian)

15. G. M. Goluzin, *Geometric theory of functions of a complex variable*, 2nd ed., "Nauka", Moscow, 1966; English transl., Amer. Math. Soc., Providence, R. I., 1969.

16. ______, *Estimates of analytic functions with bounded mean modulus*, Trudy Mat. Inst. Steklov. **18** (1946). (Russian)

17. ______, *On the Carathéodory-Fejér problem and a similar problem*, Mat. Sb. **18(60)** (1946), 213–226; reprinted as [15], Chapter IX, §7, pp. 477–492 (English pp. 497–514).

18. E. G. Gol'shteĭn, *Duality theory in mathematical programming and its applications*, "Nauka", Moscow, 1971; German transl., *Dualitätstheorie in der nichtlinearen Optimiérung und ihre Anwendung*, Akademie-Verlag, Berlin, 1975.

19. Ulf Grenander and Gabor Szegö, *Toeplitz forms and their applications*, Univ. of California Press, Los Angeles, Calif., 1958.

20. Helmut Grunsky, *Eindeutige beschränkte Funktionen in mehrfach zusammenhängenden Gebieten*. I, II, Jbuch. Deutsch. Math. Verein. **50** (1940), 230–255; **52** (1942), 118–132.

21. P. L. Duren, B. W. Romberg and A. L. Shields, *Linear functionals on H^p spaces with $0 < p < 1$*, J. Reine Angew. Math. **238** (1969), 32–60.

22. Jean Dieudonné, *Recherches sur quelques problèmes relatifs aux polynomes et aux fonctions bornées d'une variable complexe*, Ann. Sci. École Norm. Sup (3) **48** (1931), 247–358.

23. A. Zygmund, *Trigonometric series*, 2nd rev. ed., Vol. I, Cambridge Univ. Press, 1959.

24. V. Kabaĭla, *Some interpolation problems in the classes H_δ for $\delta < 1$*, Studies of Current Problems of the Theory of Functions of a Complex Variable (Fourth All-Union Conf., Moscow, 1958; A. I. Markushevich, editor), Fizmatgiz, Moscow, 1961, pp. 180–187. (Russian)

25. P. Cacridis-Theodorakopulos, *Über die untere Grenze der Rundungsschranken der beschränkten Funktionen $f(z)$, deren $|f'(0)|$ vorgegeben ist*, Math. Ann. **113** (1936/37), 657–664.

26. E. Kamke, *Das Lebesgue-Stieltjes-Integral*, Teubner, Leipzig, 1956.

27. C. Carathéodory and L. Fejér, *Über den Zusammenhang der Extremen von harmonischen Funktionen mit ihren Koeffizienten und über den Picard-Landau'schen Satz*, Rend. Circ. Mat. Palermo **32** (1911), 218–239.

28. A. N. Kochetkov, *Extremal problems with asymmetric supplementary conditions in certain classes of analytic functions*, Akad. Nauk Armyan. SSR Dokl. **41** (1965), 135–139. (Russian)

29. ______, *Extremal problems for analytic functions with positive real parts, satisfying some additional conditions*, Dokl. Akad. Nauk SSSR **148** (1963), 508–511; English transl. in Soviet Math. Dokl. **4** (1963).

30. ______, *Extremal problems with supplementary conditions for analytic functions of bounded characteristic*, Mat. Zametki **7** (1970), 63–68; English transl. in Math. Notes **7** (1970).

31. V. M. Adamyan, D. Z. Arov and M. G. Kreĭn, *Infinite Hankel matrices and the generalized problems of Carathéodory-Fejér and F. Riesz*, Funktsional. Anal. i Prilozhen. **2** (1968), no. 1, 1–19; English transl. in Functional Anal. Appl. **2** (1968).

32. ______, *Infinite Hankel matrices and the generalized problems of Carathéodory-Fejér and Schur*, Funktsional. Anal. i Prilozhen. **2** (1968), no. 4, 1–17; English transl. in Functional Anal. Appl. **2** (1968).

33. M. G. Kreĭn and A. A. Nudel'man, *The Markov moment problem and extremal problems*, "Nauka", Moscow, 1973; English transl., Amer. Math. Soc., Providence, R. I., 1977.

34. L. V. Kresnyakova, *Regular functions with bounded mean modulus*, Dokl. Akad. Nauk SSSR **147** (1962), 290–293; English transl. in Soviet Math. Dokl. **3** (1962).

35. ______, *Analytic functions with bounded mean modulus*, Izv. Vyssh. Uchebn. Zaved. Mat. **1961**, no. 1(20), 98–103. (Russian)

36. E. Landau, *Der Picard-Schottkysche Satz und die Blochsche Konstante*, S.-b. Preuss. Akad. Wiss. Berlin Phys.-Math. Kl. **1926**, 467–474.

37. Arthur E. Livingston, *The space H^p, $0 < p < 1$, is not normable*, Pacific J. Math. **3** (1953), 613–616.

38. C. J. Lewis, *The problem of Milloux for functions analytic in an open annulus*, Duke Math. J. **25** (1958), 591–600.

39. H. Milloux, *Le théorème de M. Picard, suites de fonctions holomorphes, fonctions méromorphes et fonctions entières*, J. Math. Pures Appl. (9) **3** (1924), 345–401.

40. Rolf Nevanlinna, *Über beschränkte Funktionen die in gegebenen Punkten vorgeschriebene Werte annehmen*, Ann. Acad. Sci. Fenn. Ser. A **13** (1920), No. 1 (1919).

41. ______, *Asymptotische Entwicklungen beschränkter Funktionen und das Stieltjessche Momentenproblem*, Ann. Acad. Sci. Fenn. Ser. A **18** (1922), No. 5.

42. Zeev Nehari, *Analytic functions possessing a positive real part*, Duke Math. J. **15** (1948), 1033–1042.

43. ______, *Extremal problems in the theory of bounded analytic functions*, Amer. J. Math. **73** (1951), 78–106.

44. D. J. Newman and Harold S. Shapiro, *The Taylor coefficients of inner functions*, Michigan Math. J. **9** (1962), 249–255.

45. Georg Pick, *Über die Beschränkungen analytischer Funktionen, welche durch vorgegebene Funktionswerte bewirkt werden*, Math. Ann. **77** (1915/16), 7–23.

46. ______, *Über beschränkte Funktionen mit vorgeschriebenen Wertzuordnungen*, Ann. Acad. Sci. Fenn. Ser. A **15** (1921), No. 3 (1920).

47. I. I. Privalov, *Boundary properties of analytic functions*, 2nd ed., GITTL, Moscow, 1950; German transl., VEB Deutscher Verlag Wiss., Berlin, 1956.

48. Stanisław Saks, *Theory of the integral*, 2nd rev. ed., PWN, Warsaw, 1937; reprint, Dover, New York, 1964.

49. V. M. Terpigoreva, *Extremum problems for certain classes of analytic functions with bounded mean modulus*, Dokl. Akad. Nauk SSSR **176** (1967), 1027–1030; English transl. in Soviet Math. Dokl. **8** (1967).

50. G. Ts. Tumarkin and S. Ya. Khavinson, *Qualitative properties of the solutions of certain types of extremal problems*, Studies of Current Problems of the Theory of Functions of a Complex Variable (Third All-Union Conf., Moscow, 1957; A. I. Markushevich, editor), Fizmatgiz, Moscow, 1960, pp. 77–95; French transl., Gauthier-Villars, Paris, 1962, pp. 73–92.

51. S. Ya. Khavinson, *The Carathéodory-Fejér problem for analytic functions in finitely-connected domains*, Vladimir. Gos. Ped. Inst. Uchen. Zap. **1955**, vyp. 2, 43–50. (Russian) R. Zh. Mat. **1957** #362.

52. ______, *The radii of univalence, starlikeness and convexity of a class of analytic functions in multiply connected domains*, Izv. Vyssh. Uchebn Zaved. Mat. **1958**, no. 3 (4), 233–240. (Russian)

53. ______, *The theory of extremal problems for bounded analytic functions satisfying additional conditions inside the domain*, Uspekhi Mat. Nauk **18** (1963), no. 2(110), 25–98; English transl. in Russian Math. Surveys **18** (1963).

54. ______, *On approximation, with reference to the size of the coefficients of the approximants*, Trudy Mat. Inst. Steklov. **60** (1961), 304–324; English transl. in Amer. Math. Soc. Transl. (2) **44** (1965).

55. ______, *Foundations of the theory of extremal problems for bounded analytic functions and their various generalizations*, Moskov. Inzh.-Stroit. Inst., Moscow, 1981; English transl. in this volume.

56. Maurice H. Heins, *Extremal problems for functions analytic and single-valued in a doubly-connected region*, Amer. J. Math. **62** (1940), 91–106.

57. ______, *On a problem of Walsh concerning the Hadamard three circles theorem*, Trans. Amer. Math. Soc. **55** (1944), 349–372.

58. ______, *A lemma on positive harmonic functions*, Ann. of Math. (2) **52** (1950), 568–573.

59. Dennis A. Hejhal, *Sur les problèmes linéaires extrémaux pour les fonctions analytiques*, C. R. Acad. Sci. Paris Sér. A-B **275** (1972), A329–A332.

60. Erhard Schmidt, *Über den Millouxschen Satz*, S.-b. Preuss. Akad. Wiss. Berlin Phys.-Math. Kl. **1932**, 394–401.

61. J. [I.] Schur, *Über Potenzreihen, die im Innern des Einheitskreises beschränkt sind*. I, II, J. Reine Angew. Math. **147** (1916/17), 205–232; **148** (1917/18), 122–145.

ABCDEFGHIJ – 89876